Organizational Relationships in the Networking Age

NEW HORIZONS IN MANAGEMENT

Series Editor: Cary L. Cooper, *BUPA Professor of Organizational Psychology and Health, Manchester School of Management, University of Manchester Institute of Science and Technology (UMIST), UK*

This important series makes a significant contribution to the development of management thought. This field has expanded dramatically in recent years and the series provides an invaluable forum for the publication of high quality work in management science, human resource management, organizational behaviour, marketing, management information systems, operations management, business ethics, strategic management and international management.

The main emphasis of the series is on the development and application of new original ideas. International in its approach, it will include some of the best theoretical and empirical work from both well-established researchers and the new generation of scholars.

Titles in the series include:

The Handbook of Human Resource Management Policies and Practices in Asia-Pacific Economies
Volume One
Michael Zanko

The Handbook of Human Resource Management Policies and Practices in Asia-Pacific Economies
Volume Two
Michael Zanko and Matt Ngui

Human Nature and Organization Theory
On the Economic Approach to Institutional Organization
Sigmund Wagner-Tsukamoto

Organizational Relationships in the Networking Age
The Dynamics of Identity Formation and Bonding
edited by Willem Koot, Peter Leisink and Paul Verweel

Organizational Relationships in the Networking Age

The Dynamics of Identity Formation and Bonding

Edited by

Willem Koot
Professor of Organizational Anthropology, Free University Amsterdam, the Netherlands

Peter Leisink
Professor of Public Management and Organization Studies, Utrecht University, the Netherlands

Paul Verweel
Professor of Management and Organization from a Multicultural Perspective, Utrecht University, the Netherlands

NEW HORIZONS IN MANAGEMENT

Edward Elgar
Cheltenham, UK, Northampton, MA, USA

Published by
Edward Elgar Publishing Limited
Glensanda House
Montpellier Parade
Cheltenham
Glos GL50 1UA
UK

Edward Elgar Publishing, Inc.
136 West Street
Suite 202
Northampton
Massachusetts 01060
USA

A catalogue record for this book
is available from the British Library

Library of Congress Cataloguing in Publication Data

Organizational relationships in the networking age: the dynamics
of identity formation and bonding/edited by Willem Koot,
Peter Leisink, Paul Verweel.
p. cm. (New horizons in management series)
Includes bibliographical references and index.
1. Industrial sociology. 2. Technological innovations—
Sociological aspects. 3. Communication in organizations.
4. Employees—Effect of automation on. 5. Interpersonal
relations. 6. Social change. 7. Organizational change.
I. Koot, Willem, 1940– . II. Leisink, Peter, 1952– .
III. Verweel, Paul, 1951– . IV. Series.
HD6971.O695 2003
302.3'5—dc21

2003046383

ISBN 1 84376 065 7

Printed and bound in Great Britain by MPG Books Ltd, Bodmin, Cornwall

Contents

Contributors

Jan Boessenkool is senior lecturer and researcher at the Utrecht School of Governance, Utrecht University, the Netherlands. As an organizational anthropologist he combines an interest in organizational culture, management and organization of sports clubs with organizational research in South Africa.

Marja Gastelaars is lecturer at the Utrecht School of Governance. She earned a PhD in the social studies of science and technology, and she has also taught and done research in the field of organization and management studies. She edited *On Location: The Relevance of the 'Here' and 'Now' in Organisations* (Shaker Publishing, 2000) and has contributed chapters to various books and journals.

Keith Grint is Reader in Management Studies and Fellow in Organizational Behaviour at Templeton College, the University of Oxford, UK. He is the author of numerous books, such as *The Sociology of Work* (Polity Press, 1991), *Management: A Sociological Introduction* (Polity Press, 1995) and *Fuzzy Management: Contemporary Ideas and Practices at Work* (Oxford University Press, 1997).

Marjoleine Hennis is a consultant and teaches at the department of International Relations of Trinity College-Rome, Italy. She received her PhD from the department of Political and Social Sciences at the European University Institute in Florence. Her publications include 'Globalisation and European integration, the missing link' in the *Journal of Common Market Studies* (2001), and *The Changing Role of Farmers in the CAP* (Rowman and Littlefield, forthcoming).

Paul Jeffcutt is Professor of Management Knowledge at the Centre for Creative Industry, Queen's University of Belfast, Northern Ireland. His publications include *The Management of Transition, Culture and Complex Organization in Modern Society* (Sage, 1994), *The Foundations of Management Knowledge* (Routledge, 2003) and numerous papers on organization theory.

Frans Kamsteeg is lecturer at the Department of Culture, Organization and Management at the Free University of Amsterdam, the Netherlands. His research focuses on public service non-governmental organizations in developing countries and the way in which they manage pressures towards commercialization.

Tom Keenoy is Reader in Management at the Management Centre, King's College, University of London, UK. His expertise is in discourse analysis, work organization, human resource management and employment relations. His publications include *Discourse and Organisation* (co-edited with D. Grant and C. Oswick, Sage, 1998) and 'HRM as hologram: a polemic' in *Journal of Management Studies* (1999).

Willem Koot is Professor of Organizational Anthropology at the Department of Culture, Organization and Management at the Free University of Amsterdam, the Netherlands. He has published extensively on topics such as organizational culture and theory, insecurity and management. Recent papers appeared in books edited by Sackman, *Cultural Complexity in Organizations* (Sage, 1997) and by Jeffcutt, *The Foundations of Management Knowledge* (Routledge, 2003).

Peter Leisink is Professor of Public Management and Organization Studies at the Utrecht School of Governance. His teaching and research are in the fields of work, organization and management. He (co-)edited several publications including *Work and Citizenship in the New Europe* (Edward Elgar, 1993), *The Challenges to Trade Unions in Europe* (Edward Elgar, 1996) and *Globalization and Labour Relations* (Edward Elgar, 1999).

Alfons van Marrewijk is lecturer at the Department of Culture, Organization and Management at the Free University of Amsterdam, the Netherlands. He is also a part-time senior consultant at a Dutch telecom operator. His PhD thesis *Internationalisation, Cooperation and Ethnicity in the Telecom Sector* was published in 1999 (Eburon).

David Metz is a PhD student at the Department of Organization and Industrial Sociology, Copenhagen Business School, Denmark. He is preparing a thesis on the construction of identity in 'temporary and scattered workplaces'.

Guido Ruivenkamp is Associate Professor of Technology and Agrarian Development at the Department of Social Sciences, Wageningen University, the Netherlands. He is the author of *The Introduction of Biotechnology into the Agro-industrial Chain of Production* (Jan van

Arkel, 1989) and of many articles on the social aspects of biotechnological developments.

Hans Siebers is lecturer at the Utrecht School of Governance and researcher at the CERES Research School, Utrecht University, the Netherlands. His research topics include organizational dynamics, reflexivity in organizational settings, management of diversity and globalization. 'Thinking together what falls apart: some reflections on the concept of identity' appeared in a volume edited by Driessen and Otto (Aarhus University Press, 2000).

Marcel Veenswijk is associate professor at the department of Culture, Organization and Management of the Free University of Amsterdam. He publishes on organizational culture, change processes and interventions. Recent publications are *The Dynamics of Cultural Change in Public Organisations* (Rosenberg Publishers, 2001) and 'Public-private transformations: institutional shifts, cultural changes and altering identities' in *Public Administration* (2002).

Jeroen Vermeulen lectures in organizational culture and identity at the Utrecht School of Governance, Utrecht University, the Netherlands. His research focuses on organization, multiculturalism, identity and language.

Paul Verweel is Professor of Management and Organization from a Multicultural Perspective at the Utrecht School of Governance, Utrecht University, the Netherlands. He has published widely on topics such as diversity in organizations, mergers, strategic management, organizational culture and the construction of meaning in sports organizations.

Ann Westenholz is a Professor at Copenhagen Business School, Department of Organizational and Industrial Sociology, Denmark. She has directed several international research projects funded by the Danish Social Science Research Council, of which the current one focuses on 'the social construction of identity(ies) in working life'. She is the author of various articles in the *American Behavioral Scientist* and in *Economic and Industrial Democracy*.

Ellen van Wijk teaches organization studies at Utrecht University, the Netherlands. She is also a PhD student at CERES Research School, Utrecht University, the Netherlands. She is preparing a thesis on 'commitments and organizational identification'.

Preface

In retrospect, the scientific and popular media in the 1990s reflected a widespread feeling of social discontinuities and transformations. Academics, consultants and managers were under the spell of globalization, the Information Technology Revolution, the new economy, new organization and management concepts, the new employee, new psychological contracts, and so on. Attempts to characterize the emerging social configuration resulted in concepts such as the network society. Of course there were also academics who were sceptical about the assumed discontinuities. Such academic controversy was a welcome occasion for staging a conference offering a platform for researchers interested in the ways in which organizational actors make sense of social changes and of themselves in the midst of the presumed turbulence. Thus the international conference 'Shifting Alliances and Passing Partnerships. Towards New Forms of Organizational Identification in a Fragmenting Society' took place at Utrecht, the Netherlands in November 2002.

The conference was jointly organized by the Department of Culture, Organization and Management of the Free University of Amsterdam and the Utrecht School of Governance of Utrecht University in the Netherlands, and this initiative also marked the start of their joint research programme 'The management of meaning and the meaning of management in a changing world' within the framework of CERES, the Research School for Resource Studies for Development. Some two years later, this volume brings together some of the papers that were originally presented at the conference together with a number of papers that were commissioned specifically to achieve a balanced coverage of 'organizational relationships in the networking age'. In order to examine the 'dynamics of identity formation and bonding' several levels of analysis are needed, and therefore this volume combines studies of the framing of organizational relationships at the organizational level, of personal coping strategies at the micro-level of managers and professionals, and of shifting inter-organizational relationships at the macro-level and the dynamics of identity formation and bonding in this context.

A number of people have assisted us in completing this project successfully. First of all we would like to thank Professor Arie de Ruijter, scientific director of CERES who supported our proposal to organize the conference on 'Shifting Alliances and Passing Partnerships'. Lolita van Toledo, policy official, and Agniet Cools, office manager of CERES, helped us get the conference organized and we would like to thank them for their assistance. Dr Diederik van Werven corrected the chapters written by non-native speakers and the production of the manuscript was handled by Jos Thoben. Whatever the questions that emerged during the project's progress, ranging from reflections on the title of the book to the size of the font, these were aptly and quickly answered by Francine O'Sullivan at Edward Elgar Publishing Ltd. We would like to thank her and other editorial assistants very much for their professional support. Finally we wish to express our gratitude to Edward Elgar for the confidence he showed in the quality of our project.

In spite of all the support we had, this book may have its defects, for which we accept full responsibility as a matter of course.

Willem Koot, Peter Leisink and Paul Verweel

1. Organizational Relationships in the Networking Age: An Introduction

Willem Koot, Peter Leisink and Paul Verweel

INTRODUCTION

Social processes like globalization, the information technology revolution and individualization are changing the world, and organizations, managers, employees, clients, suppliers and other stakeholders find themselves faced with the need to make sense of these changes. They will have to find ways of coping with the ensuing complexity and uncertainty of relationships in and between organizations. These phenomena are regarded by many authors as typical of the post-modern era. Theoretical explanations vary, but a number of emphases can be distinguished. Many theories have a place for developments in the socio-cultural context, which involve increasing individualization and cultural diversity of identities, norms, values and lifestyles on the one hand, and new forms of clubbing and social organization on the other hand (Maffesoli 1996). In addition, developments such as globalization are recognized which, together with the socio-cultural changes, have caused turbulence on the markets, and affect the way in which organizations operate. In response, managers have turned to strategies such as flexibilization, outsourcing, networking, teamworking, total quality management (TQM), empowerment and self-steering in order to cope with the multitude of sometimes conflicting demands. Thus, they aim to make their organizations deliver better quality service and/or to make them more cost-efficient. Another response which appears to attract increasing interest is that managers recognize that the social changes are compounded by loss of traditional commitments and collective identities such as provided by the church, trade unions and political parties, and they see a need and an opportunity for companies (and their products) to adopt the characteristics of a community with which people can identify (Casey 1995). In addition to the issue of binding and identification which tops the agenda of today's managers, self-confidence and trust are also regarded as priorities. After all, as a result of cultural diversification and fragmentation, a new foundation must be found for cooperation and mutual trust. Moreover, individuals must learn to protect themselves from increasing uncertainty and therefore learn to be self-reliant (Giddens 1990; Smets, Wels and van Loon 1999; Stacey 1996).

Theses such as these can be found in many publications. However, there are also publications which question the trends and the consequences which are presumed, such as the study by Hirst and Thompson (1996), which questions the thesis of a genuinely global economy as a qualitatively new stage in the development of international economic relationships that erodes the governance abilities of nation-states. This volume shares the aim of the attempts by the latter stream to critically examine the issues which are involved in the above sketch. The special focus of this volume is on how these social changes impact on organizations, managers, employees, clients, suppliers and other stakeholders, and how they make sense of these changes, of their worlds, of their interactions with others and of themselves, and how they (re)act to them both individually and in and through organizational relationships. In this respect, we also wish to make a distinction between their response and meaning both as professionals and as private individuals. Although the specific theories and disciplines on which the contributions to this volume are based reflect a certain degree of pluriformity, they share an interest in analysing organizational processes and the way in which these bring about organizational binding and bonding, and in the meanings which actors in organizations and organizational networks construct, and by which they manage, their relationships with others. In addition, although the specific issues on which the contributions concentrate vary, they are interconnected substantively, since they represent aspects of the overall question which this volume intends to answer, that is, how do organizational processes in the networking age and actors' sense-making interact in creating particular organizational relationships and communities? What meanings do organizational actors attribute to intra- and inter-organizational relationships in the networking age? How do they manage these relationships meaningfully, and thereby cope with the processes of social change which accompany a hybridization of organizations and pluralization of social identities which require actors' competences to navigate the ambiguities involved?

ELABORATION OF THEMES

The interaction of global social changes impacting on organizations and organizational networks, and the ways in which organizational actors make sense of these changes and processes and (re)act to them both individually and in and through organizational relationships, are at the centre of interest in this book. The paradoxical nature of these developments, organizational processes and actors' sense-making and (re)acting will be elaborated in this section.

The attempts of managers, employees and customers to find answers to the changes brought about by processes of increasing individualization and socio-cultural diversity can be understood by studying theories which examine the paradoxical consequences of the process of modernization for post- or late-modern societies (cf. Baumann 1995; Beck 1992; Giddens 1991; van der Loo and van Reijen 1997). These theories recognize that the modernization process has led to increased welfare, liberty, emancipation and greater autonomy for the individual on the one hand, but to increased risks and greater dependence because of growing specialization and global changes in the economy, science and technology on the other hand. According to Giddens (1991), we are able to grasp all the complex economic, social, legal and technological processes only partially and superficially. At best we may have a thorough knowledge of a few sub-areas, but as far as the other sub-areas are concerned we have to put our trust in the expertise of others. This entails risks. Giddens even believes that the fears connected with the awareness of the arbitrary and unpredictable character of the natural and social world around us can lead to ontological insecurity. If we are not to admit psychological defeat, we must surrender to the unknown and put our trust in particular agreements, systems, others and in ourselves. That trust in persons and in ourselves, however, does not come about readily. You have to work at it:

> trust on a personal level becomes a project, to be worked at by the parties involved, and demands the opening out of the individual to the other. Where it cannot be controlled by fixed normative codes, trust has to be won, and the means of doing this is demonstrable warmth and openness.
>
> (Giddens 1991, p. 121)

Such self-disclosure requires self-reflection, which should lead to a clear picture of one's own identity and to insight into the degree of success of one's own survival strategies, Giddens continues. In the end, this will enable self-management. Giddens is not so explicit, but one may suppose that he imagines this 'reflexive project' will increase self-confidence as well. Other analysts, on the other hand, are very explicit about this connection between openness to the other and self-confidence. According to Maffesoli (1996), openness even conveys vital force ('puissance'), and sharing these emotions with other people enhances one's own vital force even further. He thus breathes new life into Durkheim's notion of the *conscience collective* as the most primitive form of religion. In a socio-cultural respect the modernization process is leading to greater individualization but also to new forms of 'tribal' lifestyle cultures (Maffesoli 1996), both of which developments have been made possible by the demise of socio-cultural traditions.

The decline of institutions such as trade unions, which traditionally orientated workers by giving them a sense of collective ideals and identities (cf. Vilrokx 1999), has confronted individuals with the need to seek new significant orientations with the surrounding reality, and this, in turn, has led to greater complexity and uncertainty of relationships with which organizations have to come to terms as well. The flux of new management concepts, such as organizational culture, business process re-engineering (BPR), TQM and self-managed work teams is partly accounted for by their function in reducing the uncertainty of management vis-à-vis the ambiguous consequences of these social changes (Bos 2000; Grint 1997). For instance, managers attempt to create strong corporate cultures, not only because these are believed to result in company excellence (Peters and Waterman 1982), but also because strong cultures should result in integrated work communities of committed and bonded employees (Casey 1995). Through such concepts, managers also seek to appeal to employees for whom the work community can compensate the loss of traditional social institutions such as churches and trade unions. On the other hand, not only work organizations (attempt to) provide a sense of community and identity to employees and customers, but so do other organizations such as ideologically orientated NGOs like Greenpeace or leisure-orientated clubs such as a club of Harley-Davidson riders. In fact, various sorts of organizations produce their account of social changes taking place and propose new forms of social identity and action, which mediate and influence individual actor's sense-making. Theoretically, the issue is whether people are seen as social beings who need some form of community, as opposed to theories (or, better, standpoints and prescriptive models) which hold that social bonds are at best transitory and that, for instance, professional staff do not identify with the organization but with their professional career only. To the authors of these latter theories, the individual is a sort of sponge that absorbs all kinds of influences, a chameleon that can easily adjust to the most bizarre and extreme situations. His or her particularity, recognizability and predictability for others have disappeared. The individual has no continuity for him or herself. There is no clear beginning or end, now that history and the future hardly play a part any more. Only today counts. According to De Vos and Romanucci-Ross (1995), however, to have an awareness of identity a person must not only know who he or she is, but also who he or she is not. To live without a sense of continuity is 'to be faced with one's death' (ibid., p. 357).

Whatever the case, these processes have a double impact on managers. Whilst on the one hand they work hard on processes to ensure staff (and customer) loyalty, managers, themselves individuals, will seek personal coping strategies in order to survive (Koot and Sabelis 2000). For managers, just as for employees (and customers), identification with the organization is no longer taken for granted. Organizations may attempt to bond them through shares and stock options, but such economic means fall short in providing a source of identification for managers who value their activities and relationships on professional and social grounds. In fact, the term 'shifting down' indicates the recent phenomenon of young professionals and managers who give up a well-paid job in business in favour of a job in a charitable institution where they derive a sense of satisfaction from the opportunities offered by their jobs for broader social relevance and personal development. The upshot of this analysis is that work organizations should pay more attention to developing a human resource strategy which appeals to and accommodates increasingly differentiated employee needs and preferences.

It is not just socio-cultural developments, but also heightened business competition and a universal striving for efficiency and cost-reduction, leading to downsizing of organizations and outsourcing of particular jobs and functions, which impact on employees in terms of higher insecurity, more stress, higher adaptation and engagement requirements, and which contribute to individualization and increasing complexity and uncertainty of relationships. In fact, these business developments tend to dissolve the bonds that traditionally existed between employees and companies and at the same time require managers to find (new) ways of motivating employees. Strangely enough, however, these business developments have not played into the hands of traditional forms of social organization such as trade unions, which in many countries have less appeal than they used to have, despite their attempts to modernize their package of services, and to organize their activities and interactions with their members on the model of new social movements which they regard as more successful in their appeal to citizens' participation. The diversity of organizational experiences invites reflection on what bonding domains of interest individuals and organizations have in common. It also asks for reflection on how individuals in organizations, and managers and employees, in a logical sense cope with these shifting identities and social-cultural paradoxicality. How can they combine all those identities, some of which are incompatible with others, and set them in a logically meaningful context? Various analysts (for instance, Siebers 1997; Strauss and Quinn 1994) consider that there is a possible way out for individuals by not assuming a rational, serial attribution of meaning, but a parallel or associative one. In this way, there is still room for so-called impossible combinations. Incidentally, we find more or less similar ideas about the situational attribution of meaning and action in reflection on other forms of logic besides conventional Western logic, based as it is on principles like 'identification' and 'exclusion'.

The globalization or transnationalization of business does not just lead to competitive strategies. In fact, as Child and Faulkner (1998) point out, a strategy of cooperation is a counterpart to a competitive strategy, which may be intended to secure a stronger position in the market or to defend an existing position against dominant firms or new entrants. Cooperative strategy is not an alternative to competitive strategy but amounts to a further domain of policy options whose purpose is to enable firms to compete more effectively. This implies that organizations must find a balance between competition and cooperation and manage the inherent tensions. According to Child and Faulkner (1998, p. 45), trust is the key attitude that is most strongly associated with success in alliances, that is, the willingness of one party to relate to another in the belief that the other's actions will be beneficial rather than detrimental to the first party, even though this cannot be guaranteed. As in the case of the relationships between individual employees and the organization, the cooperative relationships between firms can be based on economic, cognitive or social grounds (Child and Faulkner 1998, pp. 47 ff.). Trust based on calculation of economic costs and benefits, and trust based on sharing of cognitions, including ways of thinking held in common by the parties concerned, are more usual in business transactions than in organizational cooperation based on people holding common values and sharing a personal identity. The obvious reason for the relative absence of cooperation between firms based on affect-based trust is that in business transactions some difference of interest is usually inherent in the relationship. However, according to some authors (for instance, Kotkin 1993), it is precisely because of globalization that an increase can be seen in networks formed on the basis of ethnicity, where affection plays a significant role. Such networks are thought to achieve a competitive advantage because their members trust each other blindly. Bonding between the partners, then, can develop through the individuals relating to each other and identifying with each other across the boundaries of the cooperating organizations, but this cannot sustain the relationship if either or both of the partners conclude that the calculative or cognitive basis of their cooperation has disappeared.

The management of these inter-organizational relationships constitutes the more human side of business cooperation. The complexity of managing these relationships is related to various factors such as the differences between organizational cultures, which can be manifest in the way certain goals are described and valued, and in the configuration of (in)dependency relationships between the business partners. Further complexity is added by the shape of the relationships between the cooperating business partners and governmental institutions, public and private research institutions, and other actors like consumer and environmental organizations. All actors make sense of the others' interests and actions, and try to steer these by managing the meaning attributed to their interests and actions. All these processes cumulate in an arena of relationships in which the management of inter-organizational identifications and relationships becomes a complicated matter, which makes a fascinating object of study.

SHARED ONTOLOGICAL AND CULTURAL AND THEORETICAL PREMISES

Although written by several authors representing a range of disciplines (anthropology, sociology, political theory and economics), this book does depict a high degree of common ground in terms of its ontological and theoretical premises. The following describes these shared points of departure.

Socio-cultural reality is seen as something created through interaction between individuals or groups (Berger and Luckmann 1966). People themselves 'create' a culture, which over the years acquires a certain fixed form in the sense of a normative system that leads to common habits and routines (Bourdieu 1990). In so doing, the 'creators' of the culture are indeed influenced by characteristics of the social, physical, political, technological and economic environment. However, this is not a determinant, as is sometimes explicitly and often implicitly assumed. Although cultural reality has the character of a system, it certainly does not always concern predictable, stable behaviour. Individuals bring the cultural system to life in what Giddens (1984) refers to as 'agency'. After all, they ascribe their own sense and meaning to the world around them. The logical consequence of this is that, while they can reproduce the cultural rules, they can also choose to ignore or deviate somewhat from them. They can even decide to accentuate them, use them in negotiations or for strategic purposes and 'sell' them (in a tangible and intangible sense). In other words, individuals can continually reassess what they do with this culture.

The outcome is largely determined by their perception of the situation and the people they choose to interact with, allies and adversaries alike. The political, economic and historic context within which the actors concerned place such interaction is also important. As far as potential rivals or adversaries are concerned, cultural rules are often tightened in order to create a measure of distance (Appadurai 1998; Barth 1994; Koot 1997). Elements of culture and parts of identities can also be 'borrowed' temporarily or semi-permanently and mixed or hybridized with one's own culture. The phenomenon of hybridization is becoming more common due to the processes of increasing globalization, the information technology revolution and immigration (Friedmann 1994; Werbner and Modood 1997). As a result of deepening individualization in the West and the consequent weakening of traditional structures, a higher degree of cultural fragmentation is taking place in addition to hybridization (Beck 1992; Giddens 1990): while cultures fragment further, some components regroup into new temporary configurations. Culture is therefore both continually reproduced and partially produced from the outset. This indicates a characteristic continuity and discontinuity. Hannerz (1992) draws an apt comparison with a river. It offers a degree of stability (it is a permanent feature of the landscape), but carries fresh water and can even change course in time.

Organizations are also seen in the light of two complementary perspectives: as planned and designed social systems and as providing a social environment for individuals who, although formally acting in the interests of a collective goal, strive for all manner of goals at an individual and subgroup level. Deleuze and Guattari (1987) coined the concept of a 'rhizome' which is often manifest beneath the surface: an unfathomable system of underground roots comprising informal networks and views. Thus, there is a social environment in which the same cultural and power struggles unfold, but within which there is, therefore, also concomitant cultural integration, differentiation and fragmentation (cf. Koot and Verweel 1999; Martin 2002). Partly because of more frequent mergers and the above processes of globalization, technological innovations and individualization, organizations are also no longer considered closed, goal-orientated and sustainable working communities, but are essentially seen as open, flexible networks characterized by an ongoing two-way flow of goods, cultural information and an ever-increasing number of people (Burrell and Morgan 1979). The extent to which such flows and processes of partial inclusion (Weick 1979) exist, and in connection with which organizations should indeed be seen as an endless process of identifications and divisions (Parker 1997), has served as an important theme in editing this book. This also applies to the actors within and around these organizations: do they experience the cultural complexity as sketched, and the ambiguity and multiple inclusion associated with it? And, if so, how do they deal with this? Do they only align themselves temporarily with the organization, do they experience fears and uncertainty, how do they subsequently reduce these feelings and, more importantly, how do they restructure this complex reality? The various sections of the book attempt to produce empirical answers to such questions.

In summary, in writing this book, the choice was made for a constructive, process-orientated and dynamic approach towards cultural reality, paying special attention to reciprocity between system and actor, the context, and multiplicity of perspectives and layers.

THE STRUCTURE OF THE VOLUME

Part I of this volume focuses on the way in which organizations generate identification and bonding, both as a result of intentional policies and as an outcome of organizational practices, and on how such organizational identification and bonding relate to the way in which managers and employees prefer bonds and enter into them through their actions. Bonding is, therefore, regarded as both an outcome of individual actions including the way in which individuals make sense of their actions, and an outcome of their interactions with the organization and the policies of the organization. The result may vary from intended and desired fits between individuals and organizations to unintended and undesired fits.

Boessenkool, Leisink and Verweel argue that bonds can be based on economic, political, social and cognitive relationships between employees and organizations. Current organizational practice shows clearly that organizations allude to various types of relationships when attempting to bring about bonding when labour market contingencies require them to do so. On further analysis, these relationships can be associated with two different sorts of rationalities of organizations. An organization can derive its grounds for bonding from the perspective of the organization as a goal-orientated unit and/or from the perspective of the organization as living community. The authors conclude that organizations have pursued bonding predominantly on economic and political grounds, which are dominant in the perspective of the organization as a goal-orientated unit, but they think that the process of (post-)modernization enables organizations to supplement their policies with concerns for social and cognitive bonds which are fed by a view of organizations as living communities. Such policies, however, require reflective action from all actors, because without such reflection functional rationality will remain the basis of managerial action.

Siebers and Vermeulen analyse the relationship between organization and individual actors by examining the interaction between organizational modes of coordination and the responses of organizational members. They distinguish between coordination modes based on complementarity, shared ideology, compromise, rational regulation and agreement on the basis of reasonable argumentation. Using this distinction they analyse case study material and describe the responses of actors. They conclude that identification is an ongoing, not clearly defined interactive process in which the substance and form of identification consists of negotiable and meaningful boundaries, which are dominated by the various modes of coordination.

Gastelaars analyses the relationship between individuals and organization from the point of view of the privacy which individuals may want. By using a focus that is not exactly commonplace, she is able to analyse the involvement of the individual with the organization in the processes of recruitment and marketing, selection and classification, and culturing. She takes as the starting point of her analysis that human service organizations need to adapt to processes of individualization. However, case study evidence reveals that the way in which organizations perform these processes results tendentially in the intrusion of the private sphere, stigmatization and disempowerment, harassment and intimidation. Thus these organizational intrusions upon individual lives affect three related forms of involvement, that is, symbolic involvement, political and affective involvement.

Grint questions the extent and uniqueness of the instability and complexity, which according to many authors flow from recent processes of detraditionalization and globalization, and their effect on identity formation and commitment. It is surprising through its contrast with the assumed current turbulence: the organization of commitment and identification in the Allied and German armies during the Second World War is the case which he examines. He concludes that ideological solidarity and boundary construction (us–them processes) were certainly effective, but also that these sprung from organizational considerations and commitments rather than from political ideologies (Nazi versus the free West thinking). This result is a warning against regarding instability as a characteristic of our days. Grint thinks that the culture of leadership (of the German troops) was more effective than the cult of leadership (of the

American officers). Effectiveness of leadership lies in the room it offers for initiative of lower rank organizational members (soldiers).

The relationship between organizational ideology and the commitment of organizational members is effective, according to Grint, when the culture of leadership is supported by a large degree of personal room for manoeuvre and commitment of the rank and file membership. According to Gastelaars's analysis, however, this effectiveness is permanently under pressure because of the wish of organizations to dominate the effects of individualization on organizational processes and on employees and clients. The effect is, according to Siebers and Vermeulen, that the economic and political arguments which steer the modes of coordination in organizations constantly curb the socially and cognitively orientated negotiations which the employees would want to engage in. Organizations have opportunities, according to the analysis by Boessenkool, Leisink and Verweel, to accommodate the social community rationality in organizations next to the dominant production goals-orientated rationality in organizations, and could offer more room for the social and cognitive bonds in organizations both from the point of view of organizations and from the point of view of employees. In that case, the effectiveness of organizations could – following the German leadership culture – increase with a view to both the production and the social community side of organizations.

While Part I focuses on the way in which organizations generate identification and bonding in a world that seems increasingly to diversify and fragment, Part II deals with the way in which managers and professionals in organizations deal with cultural complexity and paradoxicality. Phrased differently, it deals with ways in which individuals survive in a hectic, turbulent, ambiguous and ever faster-moving environment, where ever more (often contrasting) demands are made, not only functionally but also on private persons (which raises a separate question, namely, whether they distinguish between the two roles or whether the two overlap?).

Does the complexity produce insecurity, and in which way do people endeavour to reduce it? In which way do they cognitively cope with the chaos and complexity surrounding them? Do they still prevail, for instance, in accordance with Enlightenment ideas based on linear, causal-analytical models or are they searching for alternative cognitive approaches? What is the meaning of identity and bonding for them in a fragmenting world, and in what ways do they construct their identity? Are they searching for continuity, integrity and stability, or can they live with the ideas of changing, sometimes contrasting shared identities? The hidden question here is to what extent do managers and professionals adapt themselves in cognitive and cultural respect to the changing circumstances and why do they do so?

The first contribution by Kamsteeg and Koot directs itself to the world of Dutch top managers in commercial organizations and the world of Dutch (project) managers in Christian organizations for development aid. Both groups regard management and control as inherent parts of the job, but also as virtually impossible to achieve. The reality that confronts them cannot be 'tamed' or domesticated. The post-modern ideal that this is possible is repeatedly negated in practice. To them this means a huge amount of uncertainty. How do they deal with this, in logical and emotional terms? Eventually they seem to fall back on trust, more utilitarian forms of control and traditional methods of linear thinking. This causes emotional problems for them, because they know that these are merely sham solutions. They survive by applying strategies such as cynicism, humour, blunt execution of power and looking for support from friends, partners and professional advisers. However, they seldom act to achieve increased self-reflection and to attempt seriously to change their thinking and cultural behaviour.

This is, then, a fairly conservative, but hardly innovative strategy. When we look at a comparable project in Britain (see Watson and Harris 1999), more or less the same results become apparent. Could professionals who take another direction constitute a more renewing approach? In their contributions Keenoy, Metz and Westenholz, and van Wijk endeavour to supply an answer.

Keenoy sketches a clear picture of the way in which the world of the British academic has in the last ten years increasingly been governed by the system of research audits. These audits, held every five years, are bureaucratically complex and demanding. They constrain the academic professional, who has to adapt him or herself in terms of format, judgement criteria and time limits. Moreover, they cause strong underlying competition together with perhaps stress. Standardization, control and bureaucracy are the key words. Words that seem to be contrary to the traditional academic values of freedom and creativity. In what way, then, do academics relate to the system of audits? Perhaps somewhat unexpectedly, only some of them seem to experience this as pressure. To most it appears as simply 'the way things are' and is not perceived as untoward or problematic. Keenoy calls this attitude one of apparent instrumentality or – perhaps better – 'situated pragmatism'. British academics thus have no big problem with combining apparently contradictory values of freedom and control. His conclusion is that there does not exist one overarching coherent cultural ideal at British universities, but that academics, subject to the situation, choose this or that value. What this reveals about organizational cultures is that these are holograms rather than monoliths.

Metz and Westenholz analyse the situation, in which Danish professionals in the information technology (IT) and technological sector find themselves. Until a couple of years ago, Danish employees worked for the same organizations for long periods, and their working life was structured by their membership of a union. However, in the IT, consultancy and biotechnology sectors, the authors discern a trend towards forms of temporary and scattered communities. Employees are increasingly crossing company boundaries in their cooperation, by working as freelancers or migrants in the labour market. This means that identity construction to them is not necessarily rooted in membership of a union or a company. Identity and affiliation is not something one is/has, but something that one develops in being recognized in relation to others – in relations where recognition is 'negotiated' among involved actors in temporary and scattered work practices. Such 'negotiations' are characterized by temporary reductions in the ambiguity of 'who' the involved 'are'.

Van Wijk focuses on creative professionals in Dutch advertising agencies. These organizations have the image of being young, dynamic and trendy, and the professionals who work there are frequently regarded as persons who are fond of variation, strongly align with artistic variety and do not bind themselves for long to one specific organization. Their primary identification would preferably be with the profession and less with the organization. The two seem hard to combine. The one refers to freedom and the other to bonding. However, in the organization in which van Wijk did her case study research, the professionals (who were relatively young) seemed to be able to combine both identities fairly well. For the time that they were working there they seemed to be very loyal and committed to the organization, and when they eventually decided to leave and work as freelance professionals it was not because of a lack of organizational identification.

When we compare the results of the four contributions it is noticeable that the biggest difference appears between the managers and the professionals. The former have much more trouble in adapting themselves to the new environment, in which turbulence and complexity seem the catchwords. It may have something to do with the fact that the two are rather differently composed in demographical, sociological and anthropological respects. The professionals presented here all belong to the groups of creative spirits. In any case, they have that image. It could well be that employees in other, more traditional, sectors have more trouble with the consequences of post-modern society and organization. The managers in general seem somewhat older and are not notable for their reflexive and creative competence. Of course, the pressure of their environment is heavy on them to let it appear that they 'always have everything under control'. Moreover, in this way the environment constantly affirms the importance of instrumental-rational thinking, aimed at efficiency and linear programming.

The assumption underpinning this volume is that ongoing economic, technological and cultural transformations set a critical scene for organizational relationships and identity formation, and that processes of identity formation and organizational bonding are most salient in situations in which organizations themselves fragment and organizational boundaries get blurred. In order to examine this assumption, Part III of this volume focuses on organizations which have become part of network relationships as a consequence of the impact of a constellation of major technological breakthroughs known as information technology and biotechnology. The following questions will focus the analysis:

- Assuming that bureaucratic organizations enabled identity construction by organizational actors, what is the effect of the blurring of organizational boundaries and the emergence of network organizations for identity construction and identification with the organization?
- How do organizational actors who are involved in processes which transform organizations and organizational networks, (re)construct their sense of identities and organizational relationships?

Van Marrewijk takes up these questions in his study of a fragmenting telecom operator. In under 15 years, telecom operators have been transformed from publicly owned bureaucratic organizations into division-like structures of market-orientated business units. Van Marrewijk analyses the effects of this rapid transformation for identity construction by telecom employees and shows that the identifications by employees have not changed significantly in the sense that job and work group have remained the most salient object of identification. However, employees who were deployed by their Dutch parent company in an international strategic alliance with three foreign telecom operators had difficulties making sense of these organizational relationships and their own identity, in the ambiguity of their situation. They sensed a tension between their personal networks and bonds with their former colleagues in the Dutch parent company and the relationships with their colleagues in the strategic alliance. In the process, the former bonds turned out to be stronger as the recognition of cultural differences between the four strategic partners' employees made complex the sharing of experiences and the creation of a shared culture.

Veenswijk focuses on Dotcom organizations, which are believed to be characterized by the virtual nature of their inter-organizational network domain, multiple inclusive organizational identities and fleeting relationships, and involvement of employees. Therefore, Dotcom organizations can be regarded as an extreme case for examining the assumptions regarding the problematic nature of identity construction in ambiguous situations. Veenswijk reports on two Dotcom organizations from their inception through the recent shake-out, and the consequently changing relationships with the old economy. His research does not confirm the assumptions about Dotcom organizations. The case study organizations are not open networks, but have contacts only with a limited number of investors and clients in the old economy; neither are they virtual, since most contacts take place face to face. The images and discourses which the key actors in the Dotcom organizations construct, do not display an explosive pluralization of subjectivity but appear to follow the lines of professional group identities. His conclusion is that identity construction displays rather familiar processes of inclusion and exclusion.

The next two chapters concentrate on the biotechnology revolution which transforms the chain of seed production, agricultural production, food production and distribution, with radical consequences for the position of the actors in this chain. The contributions by Hennis and Ruivenkamp are complementary in several respects. While Hennis focuses on the way in which governments in Europe and the European Union (EU) have attempted to react to the transformation of production by a Common Agricultural Policy involving deregulation and reregulation, Ruivenkamp concentrates on the technological transformation as such, the socio-economic reorganization of the food production chain and the various possible effects on global and local levels. While Hennis explores the repositioning of farmers' organizations in relation to consumer and environmental organizations, Ruivenkamp examines the political–economic relationships in the chain, particularly those of farmers and their organizations vis-à-vis, on the one hand, the food and biochemical industry and, on the other hand, research institutes.

From a political and economic point of view, both authors pay attention to the issue of identity construction. The radical transformation of the circumstances makes a continuation of the traditional definition of self untenable or undesirable. Indeed, as Hennis shows, a differentiation is emerging. On the one hand, big farm enterprises invest in the newest technology, orientate themselves to the food industry and identify with free entrepreneurship while, on the other hand, small farmers orientate themselves to state support and new functions with regard to rural environment. This differentiation has promoted the development of diverging identities and has, indeed, resulted in the emergence of alternative farmers' organizations. Ruivenkamp argues that it is impossible for farmers and their organizations to go on organizing as sugarcane or corn farmers because the boundaries of traditional sectors fade away, or to define themselves as free entrepreneurs while they are dominated by life science enterprises. Likewise biotechnological research institutes are affected by change as the blurring of traditional distinctions between fundamental – applied and public – private research results in various new self-definitions. Ruivenkamp illustrates this by coalitions of research institutes and farmers in India and Ghana which orientate themselves to local sustainable production.

The central assumption of differentiation and fragmentation of social identities in the context of complex and ambiguous situations and fading organizational identities, seems to be confirmed by the chapters in Part III. Yet there is reason to doubt if processes of differentiation and fragmentation of identities are more radical in the post-modern era than before. The analysis of identity construction in telecom operators and Dotcom organizations would suggest this is an interactive process in work practices with the professional or social group of colleagues who are subjectively significant, and that this process is relatively independent of the turbulence confronting the organization. In addition, the contributions of Hennis and Ruivenkamp demonstrate that the way in which subjective sense-making and political–economic changes interact and may elicit each other remains an intriguing issue which continues to demand theoretical reflection.

The concluding chapter presents the main findings of this volume and concludes that these do not generally confirm the thesis of a crisis of identity and an erosion of identification with the organization, but neither do these support the idea of organizations as communities with which managers, employees, clients and others can identify and have meaningful relationships. The chapter discusses the meaning of these findings in the light of post-modernity theories and reflects on what they say about the empirical and intellectual foundations of management and organizational knowledge. It elaborates directions for future research and argues that it would be worthwhile to conceptualize culture and identity as human capacities to reframe cultural knowledge and to survive in a shifting cultural force-field. It also argues the importance of comparative research on issues of identity formation and organizational bonding in the networking age in non-Western countries. Finally, the relevance of transdisciplinary knowledge production is reflected on.

REFERENCES

Appadurai, A. (1998), *Modernity at Large. Cultural Dimensions of Globalization*, Minneapolis: University of Minnesota Press.

Barth, F. (1994), 'Enduring and emerging issues in the analysis of ethnicity', in H. Vermeulen and C. Govers (eds), *The Anthropology of Ethnicity*, Amsterdam: Het Spinhuis, pp. 11–32.

Baumann, Z. (1995), *Life in Fragments: Essays in Postmodern Morality*, Oxford: Blackwell.

Beck, U. (1992), *Risk Society. Towards a New Modernity*, London: Sage.

Berger, P. and T. Luckmann (1966), *The Social Construction of Reality*, Garden City: Double Day.

Bos, R. ten (2000), 'Fashion and utopia in management thinking', PhD thesis, Tilburg: Tilburg University.

Bourdieu, P. (1990), *The Logic of Practice*, Stanford: Stanford University Press.

Burrell, G. and G. Morgan (1979), *Sociological Paradigms and Organizational Analysis*, London: Heinemann.

Casey, C. (1995), *Work, Self and Society: After Industrialism*, London and New York: Routledge.

Child, J. and D. Faulkner (1998), *Strategies of Co-operation*, Oxford: Oxford University Press.

De Vos, G. and L. Romanucci-Ross (1995), *Ethnic Identity*, London: Sage.

Deleuze, G. and F. Guattari (1987), *A Thousand Plateaus: Capitalism and Schizofrenia*, Minneapolis: University of Minnesota Press.

Friedman, J. (1994), *Cultural Identity and Global Process*, London: Sage.

Giddens, A. (1984), *The Constitution of Society*, Cambridge: Polity Press.

Giddens, A. (1990), *The Consequences of Modernity*, Cambridge: Polity Press.

Giddens, A. (1991), *Modernity and Self-Identity*, Cambridge: Polity Press.

Grint, K. (1997), *Fuzzy Management*, Oxford: Oxford University Press.

Hannerz, U. (1992), *Cultural Complexity: Studies in the Social Organization of Meaning*, New York: Columbia University Press.

Hirst, P. and G. Thompson (1996), *Globalization in Question*, Cambridge: Polity Press.

Koot, W. (1997), 'The strategic utilization of ethnicity in contemporary organizations', in S. Sackmann (ed.), *Cultural Complexity in Organizations*, London: Sage, pp. 315–39.

Koot, W. and Y. Sabelis (2000), *Over-Leven aan de Top. Topmanagers in Complexe Tijden*, Utrecht: Lemma.

Koot, W. and P. Verweel (1999), 'Intercultureel management laatmoderne, complexe organisaties', in F. Glastra (ed.), *Organisaties en Diversiteit; Naar een contextuele benadering van intercultureel management*, Utrecht: Lemma, pp. 53–67.

Kotkin, J. (1993), *Tribes: How Race, Religion and Identity Determine Success in the New Global Economy*, New York: Random House.

Maffesoli, M. (1996), *The Time of the Tribes: The Decline of Individualism in Mass Society*, London: Sage.

Martin, J. (2002), *Organizational Culture: Mapping the Terrain*, London: Sage.

Parker, M. (1997), 'Dividing organizations and multiplying identities', in K. Hethering and R. Munro (eds), *Ideas of Difference*, Oxford: Blackwell, pp. 114–38.

Peters, T. and R. Waterman (1982), *In Search of Excellence*, New York: Harper and Row.

Siebers, H. (1997), 'Zwischen Fragmentierung und Reflexivität', *Peripherie*, **17** (67), 46–66.

Smets, P., H. Wels and J. van Loon (1999), *Trust and Cooperation: Symbolic Exchange and Moral Economies in an Age of Cultural Differentiation*, Amsterdam: Uitgeverij 't Spinhuis.

Stacey, R. (1996), *Complexity and Creativity in Organizations*, San Francisco: Berrett-Koehler Press.

Strauss, C. and N. Quinn (1994), 'A cognitive anthropology', in R. Borofsky (ed.), *Assessing Cultural Anthropology*, New York: McGraw Hill, pp. 284–97.

Van der Loo, H. and W. van Reijen (1997), *Paradoxen van Modernisering*, Bussum: Coutinho.

Vilrokx, J. (1999), 'Towards the denaturing of class relations? The political economy of the firm in global capitalism', in P. Leisink (ed.), *Globalization and Labour Relations*, Cheltenham, UK and Northampton, US: Edward Elgar, pp. 57–77.

Watson, T. and P. Harris (1999), *The Emergent Manager*, London: Sage.

Weick, K. (1979), *The Social Psychology of Organizing*, New York: Random House.

Werbner, P. and T. Modood (1997), *Debating Cultural Hybridity*, London: Zed Books.

PART I

The Framing of Organizational Relationships

2. Commitment and Community in Organizations

Jan Boessenkool, Peter Leisink and Paul Verweel

Large glass and concrete buildings – which may or may not have been of architectural value – have been replaced by small, cheerful ones with open areas, work cubicles and a few offices for top management. Staff comes and goes. Nobody any longer has a workplace of their own. People make calls, surf the net and e-mail from their study at home, their car, the garden, the restaurant and the golf course. Their territory is everywhere and they only visit the central office building for social reasons, to have a bit of a chat.
Middle management has disappeared. That carefully selected layer, built up in the period of mergers, has been shown to be superfluous because there is nobody around for them to manage any more. Top executives are in fact the only full-time inhabitants. They have been unable to adapt to more virtual forms of organizational community and they still have rooms full of works of art and personal knick-knacks intended to make the place pleasant. They continue to arrange theme sessions so as to collect staff around them, but for them, too, the most important means of communication has become the Net. Meetings have, after all, been done away with because nobody wants to waste their time taking part in exhausting marathons with low productivity. Only a few diehards show up for the Friday afternoon drinks get-together. The gatherings which the staff organise for themselves, in communities which they themselves choose, are indeed well attended, but management are not usually invited to them.
(After Munster et al. 1999, pp. 5–7)

INTRODUCTION

As this impression of the 'new times' illustrates, individual's relationships with and within organizations are subject to change. It is true that this version of a virtual organization is not yet widely prevalent, a fact which is also borne out, for example, by the small portion of the labour force doing telework. However, as organizations change through levelling of hierarchies, outsourcing of competencies, empowerment of work teams and other changes which can be understood in relation to social processes of individualization, informatization and globalization, social and organizational traditions and certainties will disappear and the power of organizations to contribute to creating an identity or to become the object of identification will diminish (Albert, Ashforth and Dutton 2000).

At the level of organizations such changes have effects which at least some managers are not convinced will be fully beneficial. As the resistance of managers towards teleworking demonstrates, the management of organizational processes is believed to be predicated upon physical presence and surveillance of employees, as well as on direct interpersonal interaction as a basis for organizational commitment. Such beliefs can be judged as emerging from instrumental concerns about organizational performance. In fact, management interest in organizational commitment, which has been accompanied by a large amount of research from the 1980s onwards, seems to spring from a belief that committed employees are prepared to work hard and to continue their relationship with their employing organization (Arnold, Cooper and Robertson 1998, p. 209). Notwithstanding this apparent instrumental interest in organizational commitment, an increasing number of organizations have attracted public interest by the emphasis they put on their organization as a community and, as such, accepting social responsibility. This latter notion, summarized by the organization accepting responsibility for people, planet and profit, involves the claim that companies do not just exist for the financial interests of their shareholders but have a broader social function including the well-being of their employees. This idea has been welcomed by governments and politicians who worry about the impact of social transformations, such as individualization, informatization and globalization, on their societies. Their worries, that these social transformations will cause the erosion of social bonds and values, could be met by civil society and private sector organizations taking up an active role in generating social cohesion.

Evidently, social transformations have an impact on work organizations themselves. For instance, there is the dismantling of the bureaucratic organization and the rise of virtual organizations, which are associated with the emergence of the network society, and which impact on managers, employees and organizational relationships (Cooper and Rousseau 1999). Yet, such concepts as empowerment and organizational culture can be regarded as good reasons for examining what opportunities work organizations provide for community and social bonds. Of course, this need for community and social bonds which people as social beings are supposed to have, can be met in forms other than through work organizations, such as clubs and social movements, but this chapter will concentrate on work organizations because these – even when they become more virtual – occupy a central place in many people's lives.

The question which this chapter will attempt to answer is twofold. First, is the recent interest of work organizations in organizational commitment and community accompanied by a new meaning of commitment and organizational relationships? Second, what could be the theoretical outline of a concept of organizational community which could offer employees an opportunity to identify with it, and enable work organizations to play an important role in creating social cohesion in the networking age?

We will first clarify our theoretical approach to organizational commitment. Introducing the central concepts, we will distinguish between various types of commitment and organizational relationships and deal with human resource policies which organizations use (or can use) to promote organizational commitment. Then we will survey some case evidence, which does not suggest that organizations have given a new meaning to commitment and organizational relationships. Therefore, we will end by discussing the issue of the organization as a working and living community, and the opportunities for creating work organizations which are not only communities for working, but also for living – since it is in such communities that human and social rationality can give organizational commitment and social bonds an entirely different meaning.

ORGANIZATIONAL COMMITMENT: A DOUBLE-EDGED SWORD

Organizational commitment is an issue both for the management of organizations and for employees. New organizational concepts reflect the desire of the management of organizations to bring about greater loyalty on the part of staff - as well as on the part of customers (an aspect of organizational strategies on which we will touch occasionally). Employees have their own commitments and needs for social bonds, which are to a greater or lesser extent focused on their work organization. These two perspectives will be elaborated.

Organizational Commitment and Management

The assertion that commitment is an issue for organizations may appear to contradict common experience. Indeed, since the 1970s, work organizations in the market sector of advanced economies have responded to technological and economic developments with a variety of different strategies, including the reduction of labour costs, increased quality of production and of services, flatter organizational structures, greater flexibility of work and organizations, outsourcing, and so on. Public sector organizations have, somewhat later, gone through similar processes, involving privatization, liberalization, budgeting, the introduction of market mechanisms and new public management. Parallel to these organizational changes, a whole range of 'new' management concepts has been promulgated, such as, for example, total quality management, human resource management, knowledge management, and cultural management. These provide managers with a range of strategies from which they can choose in order to cope with the uncertainties which confront them (Grint 1997; Karsten and van Veen 1998).

Despite management interest in benchmarking and best practices, organizations pursue different strategies. This differentiation needs to be understood partly in relation to a variety of external situations, for example the extent to which the market in which an organization operates, is confronted by economic globalization and changes in the organization's central technology. However, organizations are also capable of making strategic choices of their own accord, even if these are not unrestricted but are based on the organization's definition of the situation, its earlier choices and the constraints and opportunities these present (cf. Ghemawat 1991; Mintzberg, Ahlstrand and Lampel 1999; Pascale 1990). Thus, organizational strategies in general, and those relating to human resources in particular, are in part based on and express a certain management view of the members of the organization. This view of the employee has undergone a dynamic development, which is characterized by variation and selection (Verweel 2000). In the 1980s and 1990s, the dominant discourse focused on the rigidity of the labour market and the need for flexibilization, whereas in the course of the 1990s there came to be an increasing focus on the 'new employee' (De Korte and Bolweg 1994), a concept which referred to a high-skilled, emancipated employee who was supposed to regard employment as a business partnership. In the present discourse of the knowledge-based and networking society interest focuses on the recruitment of specific groups in the labour market (high-skilled employees as well as ethnic minorities and asylum seekers) and on organizational commitment. Management has come to realize that employees differ in all sorts of ways (age, gender, education and so on) and that they bring with them a range of motives and types of commitment to their work, meaning that organizations have to deal with the question of how this diversity can be managed.

Given the strategies which organizations have pursued (reduction of labour costs and so on), recent management interest in issues of organizational commitment and trust (see, for instance, Tyler and Kramer 1996) may raise suspicion. The creation of a contingent workforce and the implementation of such concepts as 'employability', which at least in the Anglo-Saxon context are associated with a shift of responsibility for being able to have/get a job to the worker, discourage employee loyalty to work organizations. However, rather than rejecting management interest in organizational commitment and trust generally as mere rhetoric, this interest can also be understood as a recognition on the part of management that changes in organizational practices are bound to have an impact on employee behaviour. The real issue is by what kind of interest management is driven in concrete human resource practices. Does human resource management practice express an essentially calculative conception of organizational commitment in which an effort to boost trusting relationships is undertaken because a decline in trust would entail costs? Recognizing such interest in organizational commitment and trust as representing a rational choice approach helps to critically examine human resource practices and distinguish the calculative conception of trust from what Tyler and Kramer (1996, p. 5) call a social conception of trust, in which trust has social meaning beyond rational calculations. According to this social conception of trust, people may help others or their group because they feel it is the morally appropriate action or because they identify with the group. Of course, management discourse may refer to this social conception of trust for instrumental reasons and it can attempt to allude to group or organizational identification as a means to honing employee exertion. However, it is unlikely that management can achieve such aims without any consistent action on its own part. From the management point of view this illustrates the double edge of organizational commitment: the organization cannot do without, but its pursuit will entail obligations and may be dangerous from the point of view of a calculative rational choice approach.

Employees and Organizational Commitment

An interesting point is whether the image of the staff contained within the organizational discourse fits in with the meaning which employees themselves give to their work, career and organization, and how that sense-making develops in relation to organizational strategy and practice as well as in relation to personal circumstances (life cycle) and extra-work interests. Framing the issue in these terms demonstrates an interest in a dynamic and context-sensitive approach to sense-making by employees. This involves the meaning which employees give to their work, career and organization which can evolve in relation to contexts that have little or nothing to do with the organization, such as becoming a parent or being successful in sport, and in relation to the time dimension of their occupational and life career. Thus, employee commitments to work, profession and organization and the way in which they construct their individual identity and identify with their colleagues or professional group must be understood contextually (cf. Casey 1995).

The importance of this contextuality can be illustrated by various studies which are relevant to the topic of organizational commitment. Hochschild (1997), for example, attracted attention with the results of a study which showed that American workers are working increasingly long hours so as not to have to be at home. Organizations have become more attractive as a result of new styles of management, of care and appreciation for staff, whereas the home situation has become less free because of agreements on the distribution of care tasks and the associated conflicts caused by the fact that both partners have a paid job, by the demands children make on their parents and on their time, and by the lack of appreciation for efforts at home. In contrast, the central thesis of Sennett (1998) is that – with its emphasis on flexibility and transience – the new capitalism forces career-orientated employees to hop from one job to another and constantly move house, meaning that they no longer feel at home anywhere, are unable to enter into any lasting forms of relationships, and cannot even develop any kind of coherent life story any longer. The point here is not whether Hochschild or Sennett is closer to the truth as regards employees' views – within a perspectivist view of science (Fay 1996), it is more interesting to assume two different perspectives on organizational realities. The interesting point is that the authors come up with differing significances which employees are believed to attach to work and the organization in interaction with the relevant contexts of organizations/work/capitalism and home/care/society. This underlines the conclusion of Tyler and Kramer's survey of research of trust in organizations, namely, that future research should explore the situational dimensions that contribute to shaping trust (Tyler and Kramer 1996, p. 12), or organizational commitment and community for that matter.

As in the case of management, organizational commitment has a double edge for employees. On the one hand, employees can have professional or work commitments and can identify with the group of colleagues with whom they collaborate, and such commitments and identifications need not necessarily be accompanied by organizational commitment. On the other hand, management can relate to such employee commitments and identifications, and make them instrumental for organizational purposes.

CENTRAL CONCEPTS

Commitments

The concept of organizational commitment refers to a person's attachment to the employing organization (Arnold, Cooper and Robertson 1998, p. 203). This description reflects that organizational commitment entails an affective commitment or emotional attachment to the organization. In addition to this component, Allen and Meyer (1990) distinguish normative commitment, referring to a person's felt moral obligation and responsibility to the organization, and continuance commitment, referring to a person's perception of the costs and risks associated with leaving the organization. This latter component is closely related to one of the supposed behavioural outcomes of organizational commitment, that is, staying with or leaving the current organization; as Arnold, Cooper and Robertson (1998, p. 211) observe, this behavioural outcome is, however, also dependent on labour market constraints. Another behavioural outcome attributed to organizational commitment is the willingness to exert effort on behalf of the organization.

Both affective and normative commitment refer to the employee's commitment to the 'organization', but we concur with the criticism of regarding the organization as a single entity with a united goal (Arnold, Cooper and Robertson 1998, p. 211). It is adequate to differentiate between a person's commitment to his or her colleagues or work group, to his or her supervisor and to 'the organization' meaning top management. Having observed this differentiation, the outcome of a person's commitment to his or her work group may still be counted by the management as that person staying with the organization. It would be wrong, however, to interpret this as a result of his or her identification with the values and goals of the organization, just as the person's leaving the organization could follow from causes other than a lack of organizational commitment.

Arnold, Cooper and Robertson (1998, pp. 211–12) make a distinction between the behavioural approach and a kind of social exchange approach, holding that people's commitments can be fostered by giving them positive experiences. Such positive experiences could be derived, for instance, from appreciation by managers or from intrinsic job characteristics (such as job autonomy) and extrinsic factors (including pay and other employment conditions). They report that intrinsic job characteristics are particularly important in fostering affective commitment. However, such intrinsic job characteristics are seen by other authors as a major component of another type of commitment, which is not mentioned by Arnold, Cooper and Robertson, and which could be regarded as a potential competitor of organizational commitment. This type of commitment is called 'professional commitment' by some authors (cf. Schomaker 1999) and 'career commitment' by others (for instance, Mueller, Wallace and Price 1992). Career commitment relates to the employee's identification with and involvement in his job or career. This type of commitment is primarily associated with professionals. Professionals are traditionally regarded as being primarily orientated towards their profession and, therefore, they are assumed to have a strong professional commitment and a low organizational commitment. However, the growth of interest in the emergence of 'new' professionals has led to the hypothesis that the relationship between professional and organizational commitment is contingent, depending on such factors as the type of professional and the type of organization (Schomaker 1999). Although research in this field is still fresh (cf. the chapters by van Wijk and Metz and Westenholz in this volume) one important conclusion that can be drawn, is that organizational commitment, including emotional attachment to colleagues, supervisor and/or organization as well as the intention to stay in the organization, which could result in an individual's long-time participation in an organizational community, should not be studied in isolation from other commitments and social contexts.

Types of Organizational Relationships

Following the social exchange approach to commitments, we assume that people develop commitments in interaction with others. At the level of the organization, the relationships which are produced as a result of these interactions have certain characteristics that can be more or less attuned to particular types of commitments, as can be shown by examining different types of relationships. Van Heerikhuizen and Wilterdink (1993) distinguish between economic, political, affective and cognitive relationships. These different types of relationships express the ways in which people in all sorts of communities depend on one another. Economic relationships originate from dependencies associated with the production and distribution of scarce goods; political relationships refer to those based on the compulsion which people can exert over others; affective relationships relate to processes of identity and belonging, with all the positive and negative feelings people have for one another; and cognitive relationships involve dependencies resulting from processes of acquiring and transferring knowledge. This categorization, needless to say, is an ideal one in the sense that, in actual social reality, combinations of the different types often occur and social relationships therefore can simultaneously comprise several different types of relationships.

Economic relationships

For employers and employees, an employment contract is a contract which regulates an economic exchange relationship. The employer purchases labour potential and pays for it; the employee makes his or her work capacity available for a certain period and receives payment for doing so. In this view, it is in the employer's interest to make use of the employee, and the various approaches to management which have succeeded one another in the course of time, from scientific management to human resource management, can be viewed as different attempts to achieve the maximum or optimum use of this human capital. The main concern is the effective and efficient utilization of the time which the organization has purchased. The opposite to this view, is the economic view not uncommon among some trade unions, which consider the employment contract as reflecting an unequal exchange relationship whereby the employee is exploited because he or she produces more value than that for which he or she is paid.

The relationship between the organization and the employee entails an economic exchange between two parties. From this perspective the relationship will continue as long as the employee is convinced that this relationship offers the best possible return on his 'investment' in terms of the pay and career opportunities which he receives, and as long as the employee perceives no attractive alternatives in the labour market. Allen and Meyer (1990) subsume this type of calculation under 'continuance commitment', referring to a person's perception of the costs and risks of leaving the organization. The implication of Allen and Meyer distinguishing between three components of organizational commitment seems to be that the other two components – affective and normative commitment – do not contribute to the intention to stay in (or leave) the organization. In this respect Arnold, Cooper and Robertson (1998, p. 211) offer another model of organizational commitment in which the intention to stay is the outcome of all three components and of labour market constraints. Adopting their model, leaving the organization can be understood as the outcome of a process in which economic calculation is but one consideration of reflexive monitoring.

From the perspective of management, the interest in the economic relationship is grounded in the possibilities this offers for sustaining company performance and for employee retention. Thus, for instance, offering employees stock options is based on motives to recruit valued, often highly skilled workers, to retain them, to motivate them to contribute highly to the firm and to build a sense of community, enhancing worker identification with the firm (Shperling and Rousseau 2001). However, economic means, as Shperling and Rousseau (2001, pp. 26–7) point out, will not necessarily enhance affective commitment without other practices in place that promote worker attachment to the firm, and their use entails the risk of rewarding the disloyalty of mobile workers and eroding the goodwill of workers who do not threaten to quit.

Political relationships

Organizational relationships are political relationships in the sense that these are based on the compulsion which some people can exert over others. A trend which appears to be widespread is that control through overt coercion by organizational hierarchy is being replaced (or supplemented) by control through self-managed observance of organizational policies, as for instance through various forms of direct employee participation (cf. EPOC Research Group 1997). The motives of the majority of managers, at least in the Netherlands (Huijgen and Benders 1998), for introducing direct participation is their belief that direct participation has positive effects on both company performance and quality of work, and the belief that employees have a right to participation.

Traditionally, direct participation which comes with responsible autonomy has been more prevalent among high-skilled employees, but this does not mean that political relationships and compulsion are not a feature of what would now be called knowledge-intensive organizations. In fact, one interesting explicit demonstration of compulsion exists in IT-firms where employers have used the legal construct of the competition clause to retain employees. The phenomenon of the competition clause has existed in the Netherlands since 1907 and was originally meant to prevent an employee with specific knowledge and customer relationships to leave the current firm and take a job with the firm's competitor. Lately the tight labour market occasioned IT firms to make use of the competition clause with a view to retaining their scarce IT professionals, and this has in turn motivated the government to restrict the use of the competition clause to specific situations and to a maximum of one year, with financial compensation to be paid to the employee.

Affective relationships

Interest in the importance of organizational commitment and identification arose in the early 1980s from the analysis of organizational excellence and of Japanese styles of management. Identification of the employee with the organization is believed to make the employee more committed and willing to contribute to organizational performance, because the employee feels part of the larger collective to which he or she identifies, particularly when that organization is intent on generating a sense of 'we' by focusing on the differences with competing organizations (Shperling and Rousseau 2001, p. 33). In this way organizations can become more than work organizations. A large number of historical examples have been reported of the close-knit community within and around the organization, 'modern manors' as Jacoby (1997) calls them, which offered not only employment but welfare, including company housing and recreational facilities. The community which grew up around such companies as Kodak, Sears and Thompson in the US and Philips and Shell in the Netherlands may now seem to be a total institution but, from the historical point of view, was at home in the glory days of Fordism. Nevertheless, there are examples today of companies which are more than merely places where people work, for example, the companies which provide their staff with all kinds of facilities ranging from breakfast in the 'living room' to a dry-cleaning and shopping service. The increased interest in organizational cultures has made clear the significance of such affective relationships and their influence on life within organizations.

Affective relationships do not merely involve identification with *the organization*, however. The expansion of one's sense of self to include larger sets of social entities can involve the identification of the employees not with the work organization but with *their colleagues* and with their *own class*. Various types of identification can also occur simultaneously and in conflict with one another. Guest (1987) refers to this phenomenon of double identification whereby an individual identifies with more than one organization, for example, with his own work organization and with his trade union. Identification is therefore not exclusively a matter of managerial strategy.

Cognitive relationships

Since the 1990s, there has been increasing interest in such matters as learning organizations and knowledge management, and with this comes an interest in another source of organizational commitment. Cognitive relationships within organizations can be recognized in situations in which there is, for example, a policy with respect to the training and development of employees or groups of employees. High-skilled and professional employees are believed to be particularly interested in opportunities for lifelong learning and development. Management scepticism, that organizational policies of offering such opportunities would not generate a return on investment in human capital, because those employees would leave the organization, is unfounded according to some research (for example, Gasperz and Ott 1996) because high-skilled employees would favour working with organizations which offer such development and career opportunities. Other research, however, has shown that high-skilled employees who have benefited from such human resource development policies, have a high degree of organizational commitment, but still decide to leave the organization for different reasons such as the professional 'challenge' of working at another organization or personal reasons of work–private life arrangements (Schomaker 1999).

Commitments, Identification and the Duration of Relationships

Identification in the sense of the individual expanding his sense of self to include a group of people or an organization, is associated with the idea of the individual cooperating with others and contributing to the group or organization. One effect of group identification is, as Tyler and Kramer (1996, pp. 6–7) observe, that, with identification, expectations increase that others will reciprocate. Another effect is also known, which is that people who trust others feel some sort of moral duty or commitment towards them, and are found to continue to cooperate, irrespective of the behaviour of others in the group, and not to leave groups even when it is, from an instrumental point of view, in their interest to do so. While the first effect is consistent with a rational choice explanation, according to which the individual will – other things being equal – stay with the organization as long as the organization satisfies the type of needs by which the individual is driven, the second effect would result in a long duration of organizational relationships without permanent efforts of the organization at reciprocating employee's contributions. This explains why organizations have been interested in bringing about affective and normative commitments and in 'cultural management' as a means to accomplish this.

Interest in organizational culture as an identification agent grew with the study of organizational excellence by Peters and Waterman (1982). They argue that one of the factors producing excellence is the extent to which organizations succeed in creating shared meaning. They refer to this as a strong, shared culture in which the members possess the same values and norms, which constitute a set of basic principles for thinking and acting. This set of principles functions as a point of reference for identification on the part of employees. A strong culture endows the organization and its members with the fighting spirit with which to distinguish themselves from other organizations and their members, thus creating a 'we group' – a group to which one belongs – and a 'them group' – to which one does not wish to belong. Culture thus has an integrating effect for the individual and the organization, with the individual identifying him or herself with the organization and its values and norms. Cultural management therefore focuses on providing a culture and socializing people into it. Other authors such as Martin (1992) question this approach to organizational culture because it does not take account of the diversity of people's needs and of cultural differences.

This diversity of needs has become clear in research concerning customer loyalty. In service organizations, the relationship between the organization and the customer, the contact with the customer and the supporting system for implementing that contact (including evaluating it by means of quality systems) has become the central focus of management. In a virtual economy, customers, who used to be imprisoned in a supply-orientated local economy, have potentially global and demand-driven options available to them. This 'increased' customer freedom has led management to place greater emphasis on opportunities for binding customers from within the company. Here the issue arises of customer relationship management appealing to a diversity of customers. At least four socio-cultural models for customer loyalty can be distinguished: (1) the predictable customer, who can be turned into a 'regular' customer by following the familiar rules, (2) the 'flighty' customer, who is always searching for new short-term kicks, (3) the intimate or relational customer, who constantly wishes to have the communality and permanence of the relationship confirmed, and (4) the powerless customer, who is constantly suspicious, believing that it is only lucky breaks that show that an organization, product or service provider can be trusted. Each of these customer groups is in search of a different significance with which to give substance to the type of relationship they desire, and each of them requires a different type of discourse with the organization. It is important to note that there is no direct association between the type of customer and the duration of the relationships between organization and customer, for instance, in the sense of the regular customer being more loyal than the flighty customer. Identification with the organization in categories of customers with few contacts with the organization can go along with greater continuity in the sense of returning as a customer over time, rather than within groups with large numbers of contacts.

This recognition of the importance of cultural differences is also relevant for the issue of employee retention and constitutes the basis of the second approach of cultural management (Siebers, Verweel and de Ruijter 2002). These differences may derive from functional (different departments), socio-cultural (related with ethnicity, gender or age) or political (groups within organizations) origins. This approach does not aim at the socialization of all members within a shared meaning but at coordinating the cultural differences through dialogue between the different parts of the organization, groups within the organization, and individuals. In the course of the dialogue, the differences are not eliminated but are used in the task-orientated and social sense to serve one's own purposes and those of the organization. According to this approach, identification with the organization arises when the aim of the organization makes possible the aims of the various departments, groups and individuals. Identification is offered by the organization but is also actively given substance by the members of the organization without the differences between the members being eliminated. Identification is created within and through discussion and not by socializing people into a common culture defined in advance. In other words, identification involves shared experience and not shared meaning. In the United States, this 'management of diversity' approach has become familiar primarily as an approach to organizing different ethnic groups but it has also been more broadly applied to organizational cultures (Martin 1992).

A NEW MEANING OF COMMITMENT AND ORGANIZATIONAL RELATIONSHIPS?

Having outlined the various types of employee commitments and organizational relationships, we can now address the question whether we are living in an era in which the interest of work organizations in organizational commitment and identification is accompanied by a new meaning of commitment and organizational relationships.

The interest of management magazines in the issues of organizational commitment and identification does suggest that these have taken on a new meaning for organizations in the 1990s. Survey research, for instance, of firms in the various sectors of the information and communication industry, confirms that almost all firms regard identification and the binding of their employees as very important (Leisink, Teunen and Boumans 2000). This industry includes such organizations as computer hardware and software companies, IT and management consultancies, advertising agencies, multimedia and Internet companies, audio-visual producers and graphic designers. The way in which these firms attempt to recruit and retain employees demonstrates a great variety of human resource policies, in which the appeal to various types of commitment can be discovered (cf. Computable/FNV Bondgenoten 2000; Leisink, Teunen and Boumans 2000; van Wijk 2000).

It is striking that some organizations focus mainly on the economic possibilities for organizational commitment, for example, in the form of stock options, bonuses, lease cars, WAP telephones and so on. Other organizations, in contrast, emphasize the fact that they do not wish to bind their employees by 'golden chains' but by being a community and providing a 'living space' – an informal culture, all sorts of domestic services (such as breakfast and shopping services), a bar, the freedom to play computer games and other forms of 'pleasure management' – thus promoting social bonds and emotional attachment to the organization. Finally, there are a large number of companies, which – as knowledge-intensive companies – attach a great deal of importance to knowledge management and to associated HRM policies aimed at employability and career development, elements which appeal to employees for whom career commitment is significant.

For management of organizational commitment to be effective, it is obviously essential that there should be a fit between the subjective commitments of the staff and the organizational policies. It is questionable, however, to what extent organizations are genuinely interested in their employees' commitments. Organizations in the information and communication technology (ICT) sector – particularly when the labour market was tight – were concerned mainly about the phenomenon of job hopping and losing staff, and human resource policy expressed a primarily instrumental attitude to organizational commitment. This became very evident from 2000 onwards when the Internet hype was over and the economic situation became less buoyant. An example may be taken from PricewaterhouseCoopers Consulting, which made use of various economic means and career programmes to bind their employees when it was confronted with 30 per cent employee mobility on an annual basis but which decided in early 2002 to offer financial incentives to 10 per cent of its consultants in the Netherlands for voluntary resignation. While this example of economic means to bind employees or to un-bind them can be considered as consistent human resource policy from an economic rational point of view, this view on organizational relationships would be less expected in the case of companies emphasizing that they are a community. However, one multimedia firm, Lost Boys, which represented the success of the new economy and had grown to employ over 600 employees in nine countries serving international customers like Volkswagen and Nokia, had for long worked on its image of being a community with an informal culture, but in 2001 it announced the immediate redundancy of 65 of its Amsterdam-based employees. From one day to the next these employees saw their access-card and email accounts blocked and were told that the firm would not offer redundancy pay because it was not legally obliged to do so.

It appears that the significance which organizations attach to employee loyalty is contingent on economic factors and that company turnover and profits are assessed from a shareholder rather than a stakeholder perspective. This is the case for multimedia firm Lost Boys, which made 10 per cent of its employees redundant at the same moment as its founder-owner sold Lost Boys and bought its buyer, Icon Medialab, in order to become a global industry leader. The same shareholder perspective dominates in the case of ABN-AMRO, a leading international bank whose retail bank in the Netherlands still makes a sizeable profit, although less than in the past. The bank employees are known for their loyalty, with a cashier serving 19 years of employment on average, but now the bank management is reframing this loyalty as problematic. The board of the retail bank has decided that it wants 'to take leave' of about 20 per cent of its employees before 2004 and has sent three-quarters of all employees a letter telling them that they are 'dispensable' and that no guarantee can be given about their job in the organization, which has caused much stress and demotivation among employees.

Surveying this evidence of the way in which management make sense of organizational commitment, a new meaning to organizational commitment and relationships cannot be discovered. What about the employees? Those who believe that new types of identification have arisen, refer to the development of organizational strategies such as flexibilization and employability, which are supposed to have undermined the traditional basis of employee identification with the work organization. Employees are supposed to have replaced their identification with the work organization with new forms of identification with virtual communities, which may include a virtual community of colleagues in addition to other communities (for example, Munster et al. 1999). There are examples that indicate that traditional employee loyalty, symbolized by the gold watch as a reward for 40 years of faithful service, is no longer generally appreciated. On the other hand, the supposition that informatization and individualization result in increasing numbers of flexiworkers and a trend towards virtual organizations, as a consequence of which organizations are no longer living entities to which individuals bind themselves for lengthy periods, does not get much empirical support. The proportion of flexiworkers in the Netherlands is about 10 per cent, and it is even possible to identify a slight downward trend because the tight labour market allows employees to demand a permanent contract instead of a flexicontract. This, in its turn, is associated with the fact that various studies have shown that the so-called 'new employee' (De Korte and Bolweg 1994), who is supposed to prefer flexicontracts and job hopping, constitutes a minority of those employees who are on a flexible contract; the majority prefer a stable employment relationship. The attitudes of Dutch employees towards aspects of organizational relationships do not appear to reflect a shift towards a positive evaluation of virtual communities. A recent survey showed that a majority of Dutch employees expect that competition at work will increase, that fellowship will decrease and that employee loyalty will decrease in the next two decades, but the majority also regard these developments as undesirable (Ester and Vinken 2001). This evidence does not indicate a new meaning to organizational commitment and relationships either, although the fact that a majority of employees expect future developments to be undesirable points towards the germ of an alternative concept of organizational relationships. Thus, it appears that the reason for the interest in organizational commitment and identification is the result not so much of a new meaning of organizational commitment and relationships, but of changes in management discourse in which the rhetorical pendulum has swung back from 'lean and mean' to identification and involvement.

THE EMANCIPATION OF THE ORGANIZATIONAL COMMUNITY

There is a fundamental reason, in our view, to deal with the need for a new meaning of organizational relationships. We consider that up to now the management approach to organizational commitment has been of a highly instrumental nature. The question of identification is viewed from the perspective of how the human contribution can be optimized or even maximized in order to improve the quality and quantity of production. Identification of people with one another and with the organization is predominantly at the service of the task-orientated aims of the organization. Where methods of structural reorganization (mergers, networking, business process redesign and so on) reach their limits as a means of increasing production, the interest shifts to methods of social and human re-engineering. Human resource management, organizational culture, self-management and so on then appear on the scene. However, a mechanical approach to social and human matters is ultimately dominant. This mechanization is the result of the instrumental rationalization of social life, and this appears to be the dominant perspective in theoretical studies as well. In fact, Barley, Meyer and Gash (1988) have shown by means of a statistical analysis of almost 200 articles, that the instrumental approach to culture has superseded the interpretative approach. In particular, the economic value of manipulating organizational cultures and the attention paid to the rational control of diversity by steering towards a unified culture is dominant (Barley, Meyer and Gash 1988, p. 51). We concur with this view, as our study (Verweel 2000) of the life cycle of organizational images has shown that this development actually applies to a large number of other popular images. In the initial phase of their development, various concepts – such as self-management and the learning organization – display both instrumental variants and social variants. In the following phase, however, the focus shifts towards the instrumental aspects of the concept which are directed towards organizational aims (such as increased profits and continuity) and organizational control.

Following Bahlmann and Meesters (1998) and Ritzer (1993) we think that rationalization – recognizing everything it has brought us since Taylor and Weber – has been at the cost of the intellectual and social life in organizations. The instrumental mechanistic approach enables us to tackle what is ‘tangible and manifest’ (organizational structures, decision-making and information systems, and so on) and is embedded in the enlightenment thinking, from which, according to Bahlmann and Meesters (1998, p. 28), ‘the original enlightened, humanist and social ideas have increasingly disappeared from the picture’. They argue that successive approaches have taken too little account of the fact that it is not the organization which needs to be the ultimate criterion but the individual. They conclude that the point is not merely what is tangible, but specifically what is intangible, human relationships, cultural aspects and the transcendent (ibid., p. 190).

If we wish to restrict the influence of the mechanization principle, we need to choose a different starting point than has been usual up to now. We need to re-place the organization within the development of man and society with a view to the human and social needs which organizations can serve. Viewed in this light, the essence of an organization cannot merely be contained in a business science definition, in which the effectiveness and efficiency of material production becomes the aim of the organization. Even a sociological approach in which the organization contributes, simultaneously with material production, to producing and reproducing social relationships fails to do justice to the social element, if the social element is only considered as a socialization mechanism and as a means of improving productivity and effectiveness. Then the organization is not a means to serve people's needs, but a goal in itself to which people and human functioning have been subordinated.

The analysis by the anthropologist Bailey (1977) provides an inspiring example of reconceptualizing the essence of organization. Bailey adopts the position that an organization is in fact the mixed product of the rationality of production – with its mechanistic leanings – and the rationality of the social – with its human slant – because the organization is simultaneously a community for working and for living. The working community is concerned with production and with organizing that production. These concerns are designed and evaluated according to the rationality of effectiveness and efficiency within an instrumental aim–means–output relationship. It is the prevailing economic laws of the operation of the market which dictate these processes. The living community, on the other hand, is driven by a different rationality, the essence of which can be defined as 'people treated as ends in themselves'. There is no need to make any reference to the rationality of production in order to emphasize the importance of the social and human aspects. The human aspect is, and must be allowed to be, an aim in itself. It may or may not promote productivity. The value of the organization is therefore to be found not only in increasing profits or the service provided – as the highest goal – but also in the experience and sense-making of the individual. The organization is thus the means, and the individual is the measure of things.

Thus, relationships within organizations are shaped by two types of rationality which often operate simultaneously in human action. Here, the issue of rationality posed both by Weber, Tönnies and Durkheim can be seen to return in full measure in a variant of its own. *Wertrationalität* and *Zweckrationalität* and the development of *Gemeinschaft* and *Gesellschaft*, or of mechanical and organic solidarity here enter into an entirely new relationship within organizations. Even in the digital world of e-commerce, such community features as trust, loyalty and identity are indispensable organizational principles (Reingold 1994). Old issues once more turn out to be relevant, they lead to new views of contemporary developments. In analysing how people work and live within an organization, the point is to gear the rationality of production to the rationality of the social. The point is not to reduce the social to the productive or, to put it differently, the human to the material, but to do justice to both autonomous rationalities. Organizing involves both the production of the material and of the human (Benjamin, in Visser 1998, p. 328). If we do not insist on this, we cannot avoid objectifying the subjective and thus helping to instrumentalise the life goals of mankind (Visser 1998, p. 340).

In Western society, the production of material goods has reached such a high level that any further instrumentalization, appealing to the need to provide for people's material needs, has become almost ridiculous, at least as long as the distribution of goods to the Third World and Third World areas of the West is not an issue (Castells 1996). This may open up the opportunities for advancing the rationality of the social. Another relevant phenomenon is the shift on the labour market. The demand for labour, primarily for the service sector, has increased rapidly and prospects are that this demand will increase even further over the course of the next decade, despite the economic dip in 2001–02. This means that organizations will have to persuade staff to join the organization by holding out attractive prospects to them. The need to induce staff loyalty by means of something more than just an economic relationship is making itself felt. Organizational identification by means of non-material conditions is becoming more important.

An interesting implication of interest in organizational identification inspired by social rationality concerns the issue of social cohesion. Where social cohesion has up to now been seen as the primary concern of public social policy, in future civil society and the private sector could come to play a more important role as identification with clubs and work organizations acquires increasing significance.

In short, within the actual practice of organizations in the Western world, conditions are being created (cf. De Geus 1997) or can be created which place the social relationships in organizations at the centre of attention and which create opportunities to tackle the instrumentalization of the social. In the networking age, management will be faced by a new challenge, that of enacting the concept of organization with the meaning of community of working and living.

CONCLUSION

We began this chapter with a picture of the organization in the networking age which is assumed to go along with new forms for existing types of organizational commitment. In that picture, physical proximity, an office of one's own and middle management were no longer a part of things. In our view, however, that picture – whether or not it is a realistic one – will not constitute a major break with the past. The question of what new meaning organizational commitment and relationships will acquire, needs to be answered at a much more fundamental level than that at which organizations are currently framing the question, addressing the issue of the loyalty of their personnel in order to hone organizational performance and to retain their employees as long as they wish. The fundamental issue is whether, in the late-modern era, it will be possible for organizational actors to enact a new meaning to organizational relationships, merging organizational commitment based on the mechanistic rationality of the organization of production with the individual and social need for identification with the living community and its social and human rationality. It may be that not all organizational actors make sense of organizational community in this way, but at least many employees value community, fellowship and loyalty. If such social rationality is not possible, then the form but not the meaning of organizational commitment and relationships may be able to renew itself. The current state of development of theoretical concepts and the material circumstances in Western societies suggest that such a reflexive modernization of organizational relationships is viable, but also that such a development depends on the reflexive actions of the various organizational actors, including managers and employees as well as shareholders, financiers, suppliers and customers.

REFERENCES

Albert, S., B. F. Ashforth and J. E. Dutton (2000), 'Organizational identity and identification: charting new waters and building new bridges', *Academy of Management Review*, **25** (1), 13–17.

Allen, N. J. and J. P. Meyer (1990), 'The measurement and antecedents of affective, continuance and normative commitment to the organization', *Journal of Occupational Psychology*, **63**, 11–18.

Arnold, John, Cary L. Cooper, Ivan T. Robertson (eds) (1998), *Work Psychology: Understanding Human Behaviour in the Workplace*, third edition, Harlow: Pearson Education.

Bahlmann, J. P. and B. Meesters (1998), *De organisatie die nooit bestond*, Schoonhoven: Academic Service.

Bailey, F. G. (1977), *Morality and Expediency: The Folklore of Academic Politics*, Oxford: Blackwell.

Barley, S. R., G. Meyer and D. Gash (1988), 'Cultures of culture: academics, practitioners and the pragmatics of normative control', *Administrative Science Quarterly*, **33**, 24–61.

Casey, C. (1995), *Work, Self and Society*, London: Routledge.

Castells, M. (1996), *The Rise of the Network Society*, Oxford: Blackwell.
Computable/FNV Bondgenoten (2000), *Werken in de ICT-sector: onder welke voorwaarden?*, Amsterdam: FNV Pers.
Cooper, Cary and Denise M. Rousseau (eds) (1999), *Trends in Organizational Behavior, Volume 6, The Virtual Organization*, Chichester: John Wiley.
De Geus, A. (1997), *The Living Company*, Boston, MA: Harvard Business School Press.
De Korte, A. and J. Bolweg (1994), *De nieuwe werknemer*, Assen: Van Gorcum.
EPOC Research Group (1997), *New Forms of Work Organization: Can Europe Realise its Potential?*, Luxembourg: Office for Official Publications of the European Communities.
Ester, P. and H. Vinken (2001), *Een dubbel vooruitzicht. Doembeelden en droombeelden van arbeid, zorg en vrije tijd in de 21^e eeuw*, Bussum: Coutinho.
Fay, B. (1996), *Contemporary Philosophy of Social Science: A Multicultural Approach*, Oxford: Blackwell.
Gasperz, Jeff and Marlies Ott (1996), *Management van employability*, Assen: Van Gorcum.
Ghemawat, Pankaj (1991), *Commitment: The Dynamic of Strategy*, New York: Free Press.
Grint, K. (1997), *Fuzzy Management*, Oxford: Oxford University Press.
Guest, David (1987), 'Human resource management and industrial relations', *Journal of Management Studies*, **24** (2), 503–21.
Heerikhuizen, B. van and N. Wilterdink (1993), 'Het terrein van de sociologie', in N. Wilterdink and B. Van Heerikhuizen (eds), *Samenlevingen*, Groningen: Wolters-Noordhoff, pp. 15–45.
Hochschild, A. (1997), *The Time Bind: When Work Becomes Home and Home Becomes Work*, New York: Metropolitan Books.
Huijgen, F. and J. Benders (1998), 'Het vallende kwartje; directe participatie in Nederland en Europa', *Tijdschrift voor Arbeidsvraagstukken*, **14** (2), 113–27.
Jacoby, Sanford M. (1997), *Modern Manors: Welfare Capitalism Since the New Deal*, Princeton, US: Princeton University Press.
Karsten, L. and K. van Veen (1998), *Managementconcepten in beweging: tussen feit en vluchtigheid*, Assen: Van Gorcum.
Leisink, P., J. Teunen and J. Boumans (2000), *Multimedia: de pioniersfase voorbij. Organisatiestrategie en personeelsbeleid rond multimedia in de media- en informatie-industrie*, Veenendaal: GOC.

Martin, J. (1992), *Cultures in Organizations: Three Perspectives*, New York and Oxford: Oxford University Press.

Mintzberg, H., B. Ahlstrand and J. Lampel (1999), *Op strategie-safari; een rondleiding door de wildernis van strategisch management*, Schiedam: Scriptum Management.

Mueller, C., J. Wallace and J. Price (1992), 'Employee commitment; resolving some issues', *Work and Occupations*, **19** (3), 211–36.

Munster, Ocker van, Carolien Gehrels, Max Merckx and Paul Adriaanse (1999), *De tekens van de nieuwe tijd*, 's-Gravenhage: Berenschot Fundatie/Elsevier Bedrijfsinformatie.

Pascale, Richard (1990), *Managing on the Edge: How the Smartest Companies Use Conflict to Stay Ahead*, New York: Simon and Schuster.

Peters, T. J. and R. H. Waterman (1982), *In Search of Excellence: Lessons from America's Best Run Companies*, New York: Harper and Row.

Reingold, H. (1994), *The Virtual Community*, London: Secker and Warburg.

Ritzer, G. (1993), *The McDonaldization of Society*, Thousand Oaks: Sage.

Schomaker, Pieternel A. (1999), *'Wie bindt, die wint'*, Nijmegen: Nijmegen School of Management.

Sennett, R. (1998), *The Corrosion of Character*, New York and London: Norton.

Shperling, Z. and D. M. Rousseau (2001), 'When employers share ownership with workers', in C. L. Cooper and D. M. Rousseau (eds), *Trends in Organizational Behavior*, Volume 8 Chicester: John Wiley, pp. 19–44.

Siebers, Hans, Paul Verweel and Arie de Ruijter (2002), *Management van diversiteit in arbeidsorganisaties*, Utrecht: Lemma.

Tyler, Tom R. and Roderick M. Kramer (1996), 'Whither trust?' in Roderick M. Kramer and Tom R. Tyler (eds), *Trust in Organizations: Frontiers of Theory and Research*, Thousand Oaks, London and New Delhi: Sage, pp. 1–15.

Verweel, Paul (2000), *Betekenisgeving in organisatiewetenschap*, Utrecht: Universiteit Utrecht.

Visser, G. (1998), *De druk van de beleving*, Nijmegen: SUN.

Wijk, E. van (2000), *Arbeidsbeleving bij communicatie-adviesbureaus 2000*, Amstelveen: VEA.

3. The Performative Nature of Identification: Modes of Management and Actors at Work

Hans Siebers and Jeroen Vermeulen

INTRODUCTION

Organizations are not just any kind of social entities, and organizational practices are not just any kind of practices. They exist owing to two crucial facts. First, a specific kind of practice is performed feeding the primary process of production or service delivery, which requires specific qualifications from the organizational actors. These qualifications may consist of certain knowledge or skills, derived in a rational way from the clearly defined tasks clustered together in specific function definitions. Thus these qualifications have to be checked in application procedures to see whether candidates dispose of such qualifications. These qualifications may also embody more general personality traits such as initiative and motivation, though embedded in rather open job definitions.

Second, organizational practices are influenced by policy demands, and they need some level or some extent of co-ordination. If policy would be irrelevant, and if the practices of the individual actors involved would not be geared to one another at all, no primary process of production or service delivery would be possible. On the other hand, this does not mean that organizations require every aspect of work or practice to be guided or co-ordinated. In daily work practice, there are a lot of activities individual co-workers perform on their own initiative and without the need for taking into account what other co-workers are doing. Between the extremes of absence of guidance or co-ordination and total guidance or co-ordination there is a wide spectrum. Different levels of co-ordination may come about in the course of development of the organization because co-workers gear their activities to one another, but they may also be the result of explicit management interference stemming from different modes of management. Each mode of management claims different levels of co-ordination and different requirements of individual organizational actors.

In this chapter we will approach the question of bonding from this point of view, that is, the requirements which individual organizational actors are expected to meet and the levels of co-ordination stemming from different modes of management, on the one hand, and the ways in which these individual organizational actors deal with these individual and co-ordination requirements and define and pursue their own interests, on the other hand. As a social phenomenon, bonding has to be understood in the interactions between these two elements. As a cultural phenomenon bonding is about identity construction and identification, that is, the identity formats being constructed by the management to which they expect their co-workers to answer, on the one hand, and the ways these co-workers actually deal with these identity formats and the identifications they employ in pursuing their interests, on the other hand.

Consequently, the strategic use of the identifications involved has to be looked into here. Identifications and identity constructions are very complex phenomena, and many aspects and dimensions can be distinguished. One of these constitutes the self-referential versus the performative dimensions. On the one hand, identifications and identity constructions refer to the images and narratives someone constructs about and for him or herself, in order to endow his or her life with sense and meaning and to provide structure to his or her existence and life-world and life-story. On the other hand, identifications and narratives of self play a role in interactions with others and may not reflect the self-referential dimension in an unmediated or direct way. Here the performative side comes to the fore (see Bauman 1995; 1996; Goffman 1959; 1963), emphasizing the strategic use of identifications in order to get things done, to encourage the other to do the things one would like him or her to do. It goes without saying that within the interactions between identity formats and the ways co-workers deal with them, this performative side or dimension of identification and identity construction especially needs to be studied.

In this chapter we will first discuss the various modes of management, which have emerged in the history of modern organizations. We will demonstrate the various demands these modes of management make on individual co-workers and managers, as well as the levels of co-ordination each of these management modes require. Second, on the basis of three specific cases of organizational practices, we will claim that the identity formats stemming from these modes of management are never absolute or totalizing. Individual co-workers and managers deal with them in specific ways, renegotiating boundaries and requirements, and identifying themselves in various performative ways at various moments.

MODES OF MANAGEMENT

Identity formats that management expects co-workers to comply with and the levels of co-ordination expected by the management stem from the ways the management conceives of management as a process, from the specific mode of management they apply, based on their construction of interest. For our purpose of discussing these various modes of management we might tap the vast literature on instruments of management or styles of leadership focusing on the management of organizations. Instead, we will look for more theoretical and conceptual foundations for understanding the various ways in which organizations can be managed by drawing on social sciences in general, dealing with questions of, for instance, social cohesion in society as a whole. In our view, five basic ways of managing can be derived from the contributions of the founding fathers of social sciences. These modes of management can subsequently be applied to the ways the practices of organizational actors can be managed.

In one way or another, all the founding fathers of social sciences have dealt with the problem of how to bring about social cohesion, but especially Emile Durkheim (1893) was driven by his fear of the lack of social cohesion. He epitomized this lack of cohesion in the term *anomie*, marked by a state of lawlessness and moral deficiency and confusion. He found answers to his problem in two directions. First he discussed the shift from mechanic to organic solidarity within modern society, leading to a stage in which the division of labour made individual people dependent on each other's activities. It is the baker buying meat from the butcher and the latter purchasing bread from the former. Each has an interest in the other's work. In other words, cohesion may come about in terms of practical complementarity or compatibility.

Durkheim also saw another source of social cohesion in the rise of a *conscience collective*: a culture or discourse, to use current terms, which is shared by all members of society and regulates their individual behaviour in favour of the common interest. It portrays the community or society as a socially integrated entity in which shared meanings in general or shared values in particular incite people to harmoniously live and work together. Both aspects of Durkheim's work have laid the groundwork for functionalist approaches in contemporary social sciences stressing cohesion, harmony and social integration.

Ironically, this line of thought has found support from neo-Marxist writers. Nicos Poulantzas and Louis Althusser, drawing on Antonio Gramsci, have put forward an argument similar to Durkheim's *conscience collective*, holding that class society is held together by a common ideology produced and distributed by ideological state apparatuses (Poulantzas 1975). A different current of thinking, that focuses on class and labour relations has highlighted the need for compromise to be reached in negotiations between representatives of the classes involved, of employers and employees, of management and co-workers. On the one hand, such compromise reflects common interests; on the other hand, it regulates and mediates opposing interests in an effort to meet both sides somewhere in between. The prerequisite for such compromise is that the actors involved consider themselves to be represented by legitimate institutions.

Another line of thought runs through the work of Max Weber, Karl Mannheim and Jürgen Habermas. Their terms differ, but basically they all distinguish two kinds of rationality, which we may call formal and substantial. Both kinds of rationality disclose another principle of co-ordination and the gearing together of individual actions. As Weber (1947) worked out in his famous work on bureaucracy, formal rationality emphasizes the importance of rationally designed rules, regulations and protocols, to be applied by clerks and bureaucrats in both their functioning within the organization and in their interactions with clients, customers or citizens. Obedience to, and the enforcement of, the rule of regulations is the crucial issue here in the operating of organizations. Of course, Weber was not wholeheartedly enthusiastic about formal rationality, as he saw in its application not only a bulwark against arbitrariness, corruption and nepotism, but also the danger of a consequent loss of value rationality or substantive rationality. Formal rationality would threaten the capacity of individuals to pass value judgements, to make decisions on their own based on moral standards. Consequently, social life would become inherently senseless and valueless. Kafkaesque images of individuals crushed in bureaucratic machines come to the fore here.

Habermas (1981) shares Weber's concern as the actors' life-world is threatened or colonized by instrumental or formal rationality, which is ruling the expansive system of state and market. Nevertheless, in his theory of communicative action Habermas has elaborated further on substantive rationality, which provides another device of co-ordination, pointing to joint discussion by all who are involved in a certain issue or conflict on the basis of reasonable arguments. It prescribes all actors involved to distinguish three kinds of arguments. The first kind is of a cognitive nature referring to truth claims concerning reality (to be understood in phenomenological and not positivistic terms). The second kind of arguments is about normative or moral claims regarding how reality should or is supposed to be. The third category is of an expressive kind related to how reality is experienced in emotional or aesthetic terms. It calls on those participating in the discussion to apply the appropriate criterion to each of the kind of claims: truth regarding cognitive claims, correctness in the case of the norms and moral claims, and sincerity concerning expressions and experiences. Agreement reached through discussion based on reasonable arguments in line with the appropriate criteria may be a strong co-ordinating principle for the actions of all involved.

Although Habermas does not make this distinction, his elaboration of communicative action comes close to what Toulmin (2001) defines as reason in contrast to rationality. Different from rationality, reason allows for variable and flexible definitions of reality with its validity limited to specific situations and time frames, with a feeling for the many ways in which problems can be understood, calling for modesty in style of argumentation in reaching a common definition beyond reasonable doubt, but never certain. Thus, drawing on reason may be much more adequate to solve differences and conflicts and to reach for a common definition of a situation through communicative action, rather than by rigid rationality with its timeless and universal validity claims.

To sum up, we have gathered five principles from the work of prominent social scientists, which may also guide management thinking and practice: practical complementarity and compatibility, a shared culture or ideology, institutionalized compromise, the rule of rational regulations, and agreement on reasonable arguments. To be sure, these principles and the modes of management expressing these principles reflect the basic ideas behind the ways the management tries to manage its organization. They do not necessarily reflect the actual state of affairs in the organization itself.

These basic ideas have come to be reflected in the history of modern organizations, though. The ongoing historical process of division of labour is at the heart of the development of organizations in modern societies. In a first phase, the self-sufficient or subsistence household and village economy were broken up. Households started to specialize in a limited number of economic activities and exchanged each other's products. In this phase the integrated character of the primary process of each product – each producer producing the product from the first tooling or processing to the delivery of the final product and maintaining overall overview and control – remained intact. Practical complementarity and compatibility emerged from the fact that the producers became dependent on others' products. Here the professionals were born as each category of producers – craftsmen, farmers – developed its own production qualifications to be transmitted to youngsters – apprentices, journeymen – and organizing themselves into professional associations – guilds, fraternities – to take care of common interests.

In a second phase of division of labour, the primary process was split up into a sequence of particular practices and functions, which were clustered in such a way that each cluster could be performed by one single worker. In addition, hand and mind became separated as manual labourers became the objects of managers who planned, ordered and controlled their activities. Here the foundation was laid for mass production of standardized products in industry and the standardized processing of clients on a massive scale in bureaucracies.

At first, this process of functional specialization led to a sharp decrease of qualifications each worker or clerk needed for performing his or her function as compared to the qualifications of craftsmen. This decrease even triggered deliberate attacks on such production lines: so-called Luddites destroying machines in the early nineteenth century. Within this process the need for protection of labour conditions, and especially wage interests, triggered the rise of labour associations, which were to enter into negotiations with employers.

Gradually, the rather simple functions requiring low qualifications have become displaced by machinery and technology in the subsequent waves of mechanization, automation and computerization of industry and bureaucracy. Paradoxically this led to an increase of the average knowledge and skill requirements of the remaining organizational actors in both industry and bureaucracy. Subsequently, they partly regained their overview and control over the whole primary process. Moreover, not only in industry and bureaucracy have these trends towards functional integration and the increase of training and educational demands been on the rise; the same trends are receiving impetus from the increased importance, both in economic and in employment terms, of service delivery. This means there is again a need for expanding the discretionary space of manoeuvre and relatively autonomous acting and operating of individual organizational actors.

LEVELS OF CO-ORDINATION AND DEMANDS ON CO-WORKERS

The principle of practical compatibility and complementarity can be recognized in the historical process of division of labour itself, as outlined above. It points to the fact that for actors to work together there is no need for shared meanings or culture as long as the practices of one do not contradict or obstruct the practices of someone else (compatibility), or as long as the practices involved complement each other, and all those involved are able to benefit from co-operation. For example, there is no need for similar motivations for two actors to interact and work together as long as the different motivations and intentions of both are satisfied by the interaction itself or by its result.

However, the question whether the integrated character of the primary process is maintained or not for individuals or groups of co-workers has important consequences for the level of co-ordination and the qualifications required from individual organizational actors. In case this integrated character is maintained, there is relatively little need for explicit co-ordination because the actors themselves are able to plan, design, execute and evaluate the primary process itself, either individually or as a team. At the same time a relatively high level of knowledge and skills qualifications is required here to be able to perform these tasks, compared with that needed in a primary process designed along the concept of functional specialization. Functional specialization demands high levels of co-ordination. For example, the conveyor belt requires the individual actions to be geared to one another down to the last details. Functional specialization demands relatively low skills and knowledge levels, though.

This high level of co-ordination needs in organizations designed along the lines of functional specialization has often been satisfied by the elaboration, defining, planning and implementing of rational regulations, rules and protocols prescribing individual practices in detail. It leads to workers and clerks executing these prescriptions independently of who they are, what they think and feel, and without taking into account the consequences for other co-workers, clients and customers. The rule of rational regulations entails a high level of co-ordination and allows for defining the qualifications required from individual organizational actors in strict functional terms as long as they are able to ignore substantial parts of their personality or identity. Kafkaesque images have already been referred to above. At the same time it must be stressed that the rule of rational regulations has been an important bulwark of protection against arbitrariness, nepotism, corruption and manipulation of, for example, individual workers by their bosses and individual clients and citizens by the state.

Such protection has also been provided by the principle of institutionalized compromise. This principle has already been applied by representatives of the guilds negotiating with city councils to defend their interests, for example, to keep competitors on the other side of the city walls. Industrial organizations have, of course, seen the rise of the labour movement supported or dominated by political parties in most modern societies. In the 1960s and 1970s they managed to build an extensive web of collective bargaining to defend their interests. In countries like the Netherlands this collective bargaining led to a high level of centralized agreements on even the very details of personnel management and labour relations. Since the 1990s this level of co-ordination has decreased as collective bargaining takes place on more decentralized levels and agreements leave more room for individual deals and specific personnel policies in the individual organizations. It remains to be seen whether part of this role of institutionalized compromise will perhaps be taken over by the works council.

In any case, the application of this principle of institutionalized compromise is linked to an increase of the level of co-ordination, but at the same time it guarantees at least some level of protection of co-workers' interests. The principle entails, however, a delicate package of personal qualities on the part of representatives because they have to be able to deal with different demands and play different roles, whereas of those represented a rather low level of interest calculation is required, coupled to a high level of identification, with the representatives providing the latter with the necessary legitimacy.

Associations such as guilds and trade unions have managed to develop a strong community sense and shared culture. At certain stages they even organized important aspects of the lives of their members such as insurance and funeral services. Of course, these shared cultures provided a strong basis of legitimacy for the representatives of these associations in negotiations. However, such community characteristics have also been developed by companies and other work organizations in the past. A good Dutch example is provided by the Philips company in the town of Eindhoven in the first half of the twentieth century. The Philips family and company provided housing schemes, supported educational and sports facilities and maintained close relations with the Catholic clergy to make sure that the workforce remained faithful.

Shared culture as a management principle (see Watson 1994) entails a high level of co-ordination because it means that co-workers perform co-ordinated actions by themselves because they share the meanings, which give impetus to these actions. It also means that the requirements individual co-workers have to comply with are not limited to strict functional matters.

Shared culture does not stop where functional matters end. The case of Philips demonstrates that a high level of identification with the organization as well as a stable embedding of one's own identity in the organization's culture is called for. It goes without saying that such embedding provides a strong legitimacy for management action and further tightening of co-ordination demands.

To be sure, shared meanings may also become the object of struggle for reworking and redefinition in which different groups may be involved. However, one of the problems of shared culture as a management principle stems from the fact that there are no built-in checks and balances, and there is no foundation on which fair decisions can be made. At least rational rules and regulations are able to protect all involved against arbitrariness and corruption. In the name of shared culture everything is possible, and unmediated resistance to change may harm managers and co-workers alike. For example, if the former try to implement an important plan of organizational transformation or restructuring in order to achieve a better positioning in the market, a shared culture stemming from earlier management interventions may easily stand in its way.

Moreover, because of the fact that shared culture as a management principle has no built-in foundation for critical examination and reflection based on objectivation and distantiation, there is no reason, either, for stopping at the organization's boundaries. These boundaries may not only be defined in physical terms – the gate – but also in time (separation of work time and private time) and institutional terms (system and life-world) (see Hochschild 1997). It may easily lead to the blurring of boundaries between system and life-world, instrumentalizing the latter in favour of the former. Habermas would characterize such blurring as pathological. It may provide a strong basis for ideological control of those who are best able to manipulate shared meanings, treated in a 'sacralized' way.

Nothing is sacred from the point of view of communicative action, but at the same time its scope is limited because it allows for reason to take a critical distance. Agreements concluded in negotiations on the basis of institutionalized compromise do favour a common definition of a situation which, as we have outlined above, may have far-reaching consequences for all those whose labour conditions and wages depend on such agreements. However, communicative action is not exactly what is happening in such negotiations. In the way Habermas has elaborated this concept direct participation is called for instead of representation. As a consequence, and because of the method outlined by Habermas, communicative action is especially suited for reaching agreements on specific circumscribed issues which may cause co-ordination problems. Reasonable common definitions are much more difficult to achieve in the case of general viewpoints with far-reaching (cultural) consequences. Reason emphasizes the individual's independent meaning-making and debate, and is likely to lead to common definitions only in cases of limited scope. As such it resists the totalizing claims of the shared culture principle on an individual operating in an organization and on his or her life-world. Within organizations, communicative action is geared to dynamics based on the daily functioning of the principle of practical complementarity and compatibility, to be called upon in cases when this principle turns out to be insufficient to provide for co-ordinated action.

This general analysis of the various management principles against the background of the historical development of modern organizations allows us to draw some first general conclusions. These mainly regard the levels of co-ordination required by the various management principles as well as the levels and kinds of qualifications which are demanded from individual co-workers and managers in applying these principles. This analysis suggests that the principle of practical complementarity and compatibility entails high qualifications and low co-ordination demands in the case of maintaining the integrated character of the primary process, while the opposite is true in the case of functional specialization. In both cases, however, the required qualifications can be framed mainly in functional terms. The latter also holds true for the rule of rational regulations as a management principle, which embodies high levels of co-ordination. Here qualifications demanded from managers and co-ordinators may be high, from individual co-workers relatively low.

Institutional compromise also tends towards high co-ordination levels whereas the demands on individuals are high in the case of representatives and low on represented individuals. Shared culture as a management principle has a built-in tendency towards totalizing claims regarding both co-ordination levels and the qualifications individual co-workers are supposed to answer to. After all, there is no reason why these demands should stop at functional demands. Communicative action does provide for such reasons. It favours limited co-ordination levels and as such allows for a considerable diversity in ways of working and meaning-making in those aspects which, for practical reasons, do not call for explicit co-ordination. It does require, though, that all those involved are able to distinguish between cognitive, normative and expressive claims, to apply the relevant criterion and to use reasonable arguments in discussion.

ACTORS AT WORK: CASE ONE

Let us now look at some empirical situations of organizational actors at work and consider these from the point of view of the modes of management and the levels of co-ordination discussed above. This exercise may lead to a further fine-tuning of, and elaboration upon, the general conclusions drawn above. The main point to be made here is that the inherent tension, or discrepancy, between the wish to control on the part of management and the real possibilities to achieve this in concrete situations of 'working', organizational actors strategically and situationally make use of the space that is generated by this tension. Because of this strategic and context-sensitive work, processes of identification and bonding come out as highly dynamic and unpredictable. These cases are meant to illustrate this point, and for that reason they do not necessarily need to reflect all the five management modes outlined above.

The first empirical situation to reflect upon concerns an intake interview between an administrator of a social security service and his client.[1] This situation was chosen because at first sight it may be expected that the rationality of the legal rules and regulations to be applied by the administrator will lead to highly controlled and obedient behaviour. First, let us provide some contextual information about the interview. The 'intake' takes place in a separate room at the offices, with the doors closed. This 'intake' is the first moment of face-to-face interaction between the client and the administrator. The client is claiming a social benefit, because at that moment he is jobless. He left school in order to marry his girlfriend. She, now his wife, is also present at the meeting. During the intake, along with some formal information, the client's right to social benefit has to be assessed by the administrator. In his talk with the client, the administrator argues that both the man and the woman are too young to be granted the social benefit. The only way to get any money is to ask their parents to give them the child allowance. Of interest can be the fact that both the man and the woman are second-generation Turkish immigrants in the Netherlands. The social meaning of marriage is probably different, especially when it comes to the role of the man.

Administrator:	I consider you as two different persons, the law considers you as two different persons

The administrator states that their age is the most significant feature of their social situation, and explicitly not their marriage. He supports this by referring to the law. The exact phrasing is of interest here: it is characterized by two syntactically identical phrases in which 'I consider you ... ' is substituted by 'the law considers you ... '. This parallelism shows the strict identification with the rules of the law, typical for bureaucracy. The interactional purpose obviously is to deny the possibility of contesting the point of view of the administrator, to prevent any possible reasonable dispute about the interpretation of the rules and regulations of the bureaucracy. It may even serve to prevent a dispute about the fact that the statement of the administrator is a subjective interpretation. His exclusive knowledge of the law qualifies him as the professional. His professionalism is thus based on an apparent strong identification with the legal and organizational context. Identification with the personal context of the client, from the perspective of the law and the organization of social security, is to be regarded as more or less unprofessional behaviour:

Client:	If you are in my situation, hey, if you are in my situation, what should I do?
Administrator:	I am not in your situation.
Client:	Yes, if you are in my situation.
Administrator:	I am not, I am not.
Client:	No.
Administrator:	That's why I don't want to go into it, I am not in your situation.
Client:	No, I'm just giving an example.

The administrator vehemently refuses to see the case from the point of view of the client by using the same words and phrase 'I'm not in your situation' three times. His refusal to change viewpoints creates a kind of 'cul de sac' conversation with the client. As in the first fragment of the interview, this short exchange shows that the work of bureaucracy is not at all based on Habermasian reasonable communication. The rationality of rule-application excludes personal empathy and identification with the client. But, as appears from the insistence of the administrator, rationality has to be defended almost emotionally. This can be seen as evidence of the fact that the administrator is performing his role and that he has a strategically motivated interactional goal by acting this way. By distancing himself from the client's situation he is creating the space for applying 'the law' on this specific case, not (only) for the law's sake, but in order to help the client. In the following exchange, he admits their problematic situation and explains his strategic behaviour vis-à-vis the law. He now reverses, one can say, his identification: coming closer to the client and taking some distance from the law.

Administrator:	I am going to interrupt you for a moment. In itself, uhm, I hear you say that you are in a difficult situation, I can understand that.

Client: Yes!

Administrator: On the other hand, there is legislation and legislation is bounded by certain rules and if you don't conform to these, then I am not able to do anything for you despite the fact that you are in a very difficult situation.

And

Administrator: Hey, I am giving you advice on which grounds you might receive a social benefit and when not.

On the one hand the law seems to stand as a sort of objective fact or beacon in the turbulence of the actual conversation and the actual circumstances. There is a strong identification on the part of the administrator. On the other hand, the law is mainly used strategically by the administrator: to give himself an authoritative position in the interaction, to delimit the space for the client and to look for solutions that can help the client within these limitations. The boundaries of the level of co-ordination, entailed by the law, are narrow. Space to act, to interpret and apply the law, is presupposed. There is no option for the administrator but to act in the space between the letter and the application in specific situations. This means that, even in a bureaucracy where rationality rules, the level of co-ordination is never total. Hence, the institutionalization of the professional's space to act in the term 'discretionary space'. The organizational term 'discretionary space' then can be seen as a *post hoc* legitimation by management of the freedom a professional worker always has to take.

In terms of qualifications of the worker, this means that on an explicit level he or she must have knowledge of the rules of the law. But on a more implicit level he or she has to be able, in applying the law, to perform different styles of speaking and acting, implying different identifications. In other words, more is involved than just functional qualifications as we suggested above. The level of co-ordination may be relatively high, but apparently the rule of rational regulations can only 'work' due to an important discretionary space granted to individual administrators as well as the ability to apply various identifications at various moments.

ACTORS AT WORK: CASE TWO

The second empirical situation shows organizational actors at work in a professional organization (a college), where the level of co-ordination is much lower and more contested.[2] The setting considered here is a meeting between a team of teachers from the department of Theology and their operational manager. The level of co-ordination is more vague because, on the one hand, the operational manager, Piet, is a professional himself (a teacher of pedagogy) and, on the other hand, the chairman of this meeting, Wim, the head of the department, is both a professional and a manager. The meeting has been organized because of a letter sent by the head of the department to the manager in which he complains against the effects of the re-structuring of the organization. According to the letter the re-structuring will be detrimental to the quality of the education programme of the department and affect the time and care the teachers can provide for the students. The tone of the letter is quite offensive and refers to Paolo Freire's concept of the culture of silence.

Excerpt 1[3]

1. Piet:	What I **absolutely disagree on and I want to state this emphatically is the survival-culture and the quotation from Paolo Freire, because Paolo Freire wrote this in an book on the pedagogy of the oppressed. And, well, this comparison [laugh] I *don't agree with, not in any sense...no. That's the only thing I want to say about it ...
2. Greet:	Can you explain this a bit further?
3. Piet:	That the culture of silence belongs in South America and constitutes an established culture there.
4. Greet:	Okay, but beside this quotation, the excerpt above it, that is the reason for it, do you disagree on this as well?
5. Piet:	That there is no attention for people at this school, I disagree on that too yes. That emphasis is at present a bit more on cutting expenses and on control – I can't really say this by the way – but on reducing expenses yes that can't be denied. But that attention for the people is absent that uhm I disagree with. But that is later, that can be reacted on fully, of course...
	[About two minutes later at the end of his monologue:]
	And what I want to say about it furthermore in general, such a letter doesn't come out of the blue. That I realize very well. And how it will be next, and how it is worded I don't care so much if we can work it out together with reference to what we expect from each other. If we can work this out this afternoon, I'll have to leave at four o'clock, ... then please.
6. Wim:	Thank you very much.
7. Piet:	Yes?

The reference to Paolo Freire is rejected fiercely by the manager, Piet. He does not want to be associated with it, and he does not even want to discuss it. It is obvious that the writer of this letter explicitly speaks to him as a professional colleague, being fully aware that the manager – as a pedagogue – knows the text of Freire well. The writer – the head of the department – tries to establish common ground in this way. The manager cannot but refuse to go into this, because it would undermine his authority as a superior – as a manager. And on the level of content, it is also impossible for him to discuss it (and thereby treat it as a legitimate argument of the present discussion) because Freire's phrase directly refers to power relations and the revolutionary rejection of these relations.

This first exchange shows immediately how fragile the co-ordination here is, exactly because of the possibility of identifying on the level of professional colleagues and the vagueness of the hierarchic relations. There is no 'objective' point of reference, such as the law in the first situation, for identification. Because of the situatedness of the identification, the boundaries have to be constructed ad hoc in terms of place, time and activity. In this meeting the manager does not want to be addressed as a professional, he sets a time (four o'clock) and an interactional goal ('to establish expectations'). The boundaries have to be established, for they are not given in the situation. They are potentially very dangerous for undermining the level of co-ordination between the organizational actors.

Excerpt 2

1. Wim (chairman): On the other hand, I'm watching the time now, it's ten minutes to four o'clock and we will have to speak out expectations. Okay, Feiko, do you want to say something? <5> Well, as far as I am concerned I take over Piet's opening question, to state what we * want together, what we * expect, where we want to go to.

This fragment shows the identification of the head of the department Wim with the manager. Without discussing the goal of the meeting, he just takes the manager's statement of the meeting's purpose 'talking about expectations' as the legitimate one. Notice that the wording of this goal is personal, aims at a kind of psychological contract between the actors present and has no consequences for action on the part of the management.

Excerpt 3

1. Wim (chairman):	I am watching the time ... I'm asking everybody to just speak out expectations.
2. Dick (teacher):	I think that the students, who are not represented here, have certain expectations about the study programme and that these have been put forward already. And, actually, I would like to get a reaction on that. I didn't hear much reaction on that.
3. Piet (manager):	With reference to the falling student numbers?
4. Dick:	[Interrupts]: The reasons why students have the feeling that this is not their home any more and the concern we have about it. Really, it goes to our heart. Yes, the feeling of, gee, let's do something about it together and what can we expect from it, about what is going to happen.
5. Piet:	Yes, but speak out an expectation toward management please.
6. Dick:	Yes, I would like very much that the (lesson unit) would be enlarged, that there will be space for it. And, to myself, what is your concern. At a certain time you are being squeezed. I will give an example. If you give 24 modules a year in the restricted establishment that you get for it. At a certain moment it points to running your own store and yes, to keep one's head above water. ... I would like to have some more space to give lessons of course.
	[He continues for 1.5 minutes]
9. Wim:	That's your question, time for lessons.
10. Piet:	Lesson, unit, space.

Again the head of the department repeats the goal of the meeting. The teacher Dick tries to get a discussion with the manager about the negative effects for the students and his own work of the re-structuring of the organization. The manager in 5 reacts irritatedly, because the teacher does not comply with the goal set for this meeting 'speaking about expectations'. The elliptical, non-understandable sentence, of the manager in 10 underlines his lack of interest and the total absence of any reasonable communication.

In short, we are dealing here with a case of practical complementarity in which the primary process maintains a large part of its integrated character. In the end each of the professional teachers determines the content and process of the events in the classrooms. This example confirms that such cases may not require a high level of co-ordination, but the ways to reach any co-ordination that is required, for instance, regarding time frames of the classes and cutting back in expenditure, may be very difficult. The grounds on which to achieve such co-ordination may be shaky. For the managers concerned the option to identify themselves in clear terms, as the social security administrator was able to do initially, as such providing a basis for authority, is hardly viable because they themselves are also identified as professionals, in similar terms to those whom they are supposed to 'manage'. Professional identifications are even used in political terms when the revolutionary connotations of the vocabulary of Paolo Freire is evoked by those professionals.

In this example the participants in the meeting mainly resort to discursive manoeuvres to make their points. The managers invite the participants to express their 'expectations', not their 'arguments' or the 'reasons' for the points they would like to make. Expectations can be gathered and expressed allowing the managers to stick to such an inventory, whereas arguments and reasons unavoidably call for discussion and the construction of a common definition of the situation. The latter would call for communicative action. Here discursive struggle takes the place of reasonable discussion, which is expressed in the fact that rhetorically and in terms of linguistic logic Dick finds it extremely difficult to express himself. A similar discursive manoeuvre is used by manager, Piet, who ends the meeting with words without any meaning or logic whatsoever.

In other words, professionals involved in such cases in which the integral character of the primary process *grosso modo* remains intact must not only answer to high qualifications demands regarding their profession. This example shows that they must also master the rules of the power games involved, for instance, discursive manoeuvring, in order to make their points. It is remarkable, however, that although they are all academically trained, they resort so little to reasoning based on arguments to arrive at a common definition of the situation in a communicative way.

ACTORS AT WORK: CASE THREE

The meeting between the professionals of the department of Theology and their manager ended at four o'clock, the time boundary set by the manager. After the manager left the meeting, the discussion among the professionals continued. Teacher, Feiko, who was explicitly invited earlier during the meeting (by the chairman, see Excerpt 2) to make a contribution but silently refused to do so, almost immediately starts talking.

Excerpt 4

1. Feiko (teacher):	I found myself in a difficult position so I kept silent as much as possible, because actually I don't agree with you at *all I noticed.
2. Greet (teacher):	Hm.
3. Feiko:	I do have a totally different image. I think you make a very cramped and conservative impression.
4. Greet:	Hmhm.
5. Feiko:	And it surprises me that the management team doesn't give you tit for tat about what they hear here. I think they just will ...
6. Greet:	Explain.
7. Geert (teacher):	Really, give some *good explanation.
8. Greet:	Yes.
9. Geert:	I will listen well to it [turns his chair in the direction of Feiko].
10. Feiko:	Yes for instance I think you complain that the students complain that is something that we as teachers created by evoking totally wrong expectations. In which they get disappointed accordingly.
11. Greet:	Such as? What kind of expectations?
12. Feiko:	For example the warm nest as it is called here, or whatever should be a warm nest. If you don't suggest such a thing in the lessons, that it could exist, then no student will expect to get it. ... Then it will be a much more business-like study programme.

Feiko starts criticizing his colleagues for holding on like grim death to their old ideal of education, thereby disengaging from their standpoint in the 'struggle' with management. It is remarkable that he does so only after the manager left the meeting. Had he done this in the presence of the manager, the effects in terms of identification would have been substantial. In that case his words would have had the effect of identifying exactly with the manager against his own colleagues. Now, by not talking in the presence of the manager, he has shown solidarity with the views of his fellow-professionals. In fact, he has not accepted the situationally constructed boundaries set by the manager, the four o'clock and the goal of the meeting (expressing expectations). By waiting to speak he has strategically created space to make a contribution that has at least the potential of generating a discussion on the issue and on the existing different opinions. This discussion was strikingly absent during the first hour of the meeting, which was characterized by ritually moving positions after the refusal of the manager to create common professional ground by accepting Freire's terms. Feiko managed to remove himself from the imprisonment in which the other actors kept themselves by agreeing with the boundaries placed by the manager. It allows him to be critical about the points of view of the professionals, but also about the behaviour of the manager (whom he criticizes for being inactive towards the professionals). His words (and his timing) show painfully how powerless both parties were during the first part of the meeting. A minimal level of co-ordination could only be preserved by agreeing on the, just as minimal, terms set by the manager. One could argue whether the level and kind of co-ordination between the professionals and the manager can be 'workable'. Is it not just 'keeping up appearances', typical front-stage behaviour, growing out of subscribing to the ideology of shared culture as the only basis for co-ordination in a professional organization? The fear of (discussing) differences is bigger than the need to work on solutions for the existing dilemmas. But, does this really matter here? Is it not true that the powerlessness of the functional relations and the lack of substantial co-ordination can easily be overcome by just acting in situations where one does not have to confront one another? As Feiko shows, space for acting strategically is found simply at another time and place.

CONCLUSIONS

We began by defining and describing five modes of management, arguing that organizational bonding is created in the tension between these management modes, on the one hand, and the ways workers relate to them in specific practices, on the other. Management modes can be seen as identity formats and the actual practices could be defined as settings where the play of identifications is staged. We have first demonstrated that different management modes entail different claims on co-workers and managers in terms of qualifications and imply different levels of co-ordination, in other words lead to different identity formats. Second, we have tried to analyse in our examples of 'actors at work' how the boundaries set by these organizational formats are endowed with meaning, become meaningful.

In social terms the relational aspect of bonding comes to the fore in the ways that all those involved relate to one another. To what extent are they able to pursue their interests and improve their positions within these relations? On the one hand, modes of management clearly set the limits of their space of manoeuvre. The social security administrator cannot simply decide not to apply the laws as prescribed by the rule of rational regulations. His clients also cannot do anything but to accept these rules. The participants in case two cannot afford any attempt to step out of the consensus straitjacket required by the shared culture mode. This case also shows, however, that the ground for any attempt to co-ordinate and manage is shaky when the integrated character of the professionals' work is maintained.

On the other hand, the imperatives stemming from these modes are not so pressing as to exclude all initiative and agency on the part of the individuals concerned. The social security administrator is able to do something for his clients. In the end he manages to please both them and the ones to whom he is supposed to account for his decisions using his discretionary space. Feiko in case three is able to turn the ingredients of the situation, even those instruments deployed by the managers, to his own advantage.

As these cases demonstrate, boundaries are never fixed nor total. Actors at work are able to use them strategically, thereby showing that their identification 'work' is situationally defined as well. Even in a situation where the boundaries seem stalemate and fixed by 'the law', as in the case of the social security administrator, and consequently the distance between the worker and his client seems unbridgeable, a good worker can create space to deal with the rational rules in a more human way. Within this space reason is not completely ruled out by rationality in the interaction between client and worker.

Nevertheless, the social security administrator is certainly making use of his instrumental-rational or calculative faculties in order to create or safeguard his space for manoeuvre. It must be stressed, though, that successful manoeuvring on the part of each individual actor within the existing framework of power relations not only requires such rational faculties to bring into the power game, it also requires other cultural capabilities such as taking the lead in verbal encounters and in defining the situation, in bestowing meaning on both text and context. It even involves the effective deployment of one's own identification repertoires. One's own performed identifications play a crucial role in creating and safeguarding one's space of manoeuvre within existing identity formats, as all three cases demonstrate. Here the social and cultural aspects of bonding merge.

However, meaning-making can also become solidified or petrified when existing meanings and conceptions are taken for granted as a guiding management principle. As case two indicates, the meeting between manager and professionals seems to be a cul-de-sac, from a communicative point of view. Management-by-shared-culture seems to imprison the actors in a destructive hold: no one is really able or prepared to break the code of 'shared culture' and talk plainly about the differences. Especially the manager refuses to define a framework in which the differences can be talked about. He sets an agenda (talking about expectations) which is meaningless, and keeps to a time boundary. These two minimal boundaries keep the communication going ritually for one hour and all seem to be trapped in a helpless way by the implicit consensus standards developed and set by the organization.

There seems to be no escape but, as case three shows, there is a way out by choosing a different time to talk about the issues. Simple as this may sound, this is exactly what organizations are about: more or less co-ordinated place and time networks of actors at work. To be able to use this in a strategic manner is a way to make bondings and identifications dynamic and situationally defined within negotiated and meaningful boundaries set by the relevant mode of management.

Thus, bonding is not something which can be understood as a static or wholehearted identification of individual actors with the organization, including a stable self-referential component. Instead, bonding has a conditional and relational character in which the strategic and performative aspects come to the fore. Bonding is particularly about performance and strategy based on calculated and constructed interests, rather than about any stable or unmediated self-referential process. In social terms bonding is about power, in cultural terms it is about meanings and identifications being used to reach certain objectives. In our view, an attempt to revive a romantic portrayal of bonding as a self-referential identification with the organization in trendy terms of 'commitment', ' loyalty' or 'motivation' discloses a clear ideological character by ignoring these social and cultural dimensions.

NOTES

1. The data presented in case one are the result of an empirical case study of a social service organization in one of the major cities in the Netherlands, carried out in 1995. The data have been translated into English, the original language being Dutch.
2. The data presented in cases two and three are part of dissertation research carried out in 1997 (see Vermeulen, 2001). The data have been translated into English, the original language being Dutch.
3. Transcription conventions used: silences are marked by ... for a relatively short pause (less than a second) and by < > with number for a specific long pause, for instance <3> for 3 seconds; round brackets without text ( ) indicates unintelligible speech, while round brackets with text indicates probable speech; a (double) asterisk * indicates emphasis.

REFERENCES

Bauman, Z. (1995), *Life in Fragments: Essays in Postmodern Morality*, Oxford: Blackwell.

Bauman, Z. (1996), 'From pilgrim to tourist – or a short history of identity', in S. Hall and P. Du Gay (eds), *Questions of Cultural Identity*, London, Thousand Oaks, New Delhi: Sage, pp. 18–36.

Durkheim, E. (1893), *De la division du travail social*, Paris: Alcan.

Goffman, E. (1959), *The Presentation of Self in Everyday Life*, New York: Doubleday.

Goffman, E. (1963), *Behavior in Public Places: Notes on the Social Organization of Gathering*, Glencoe: Erving Free Press.

Habermas, J. (1981), *Theorie des kommunikatieven Handelns*, Frankfurt-am-Main: Suhrkamp.

Hochschild, A. R (1997), *The Time Bind: When Work Becomes Home and Home Becomes Work*, New York: Metropolitan Books.

Poulantzas, N. (1975), *Classes in Contemporary Capitalism*, London: New Left Books.

Toulmin, S. (2001), *Return to Reason*, Cambridge, US: Harvard University Press.

Vermeulen, J. (2001), *De naam van de school*, Baarn: Ten Have.

Watson, T. (1994), *In Search of Management: Culture, Chaos and Control in Managerial Work*, London/New York: Routledge.

Weber, M. (1947), 'Bureaucracy', in H. H. Gerth and C. Wright Mills (eds), *From Max Weber: Essays in Sociology*, London: Kegan Paul, pp. 196–244.

4. The Critical Edge of Individualization: About the 'Culturing' of Service Delivery and the Performance of Privacy

Marja Gastelaars

INTRODUCTION

Modern-day organizations, particularly organizations that highlight the individuality of both their clients and their employees, are subject to a familiar paradox of organizational life. On the one hand, one may expect that their professed interest in their clients' expectations and experiences, and in the lives and careers of their employees, should also be beneficial to those individuals' organizational involvement. It is this explicit catering to individual experiences, preferences, and particularly to the symbolic values associated with the organization, that I call, after Du Gay and others, the 'culturing' of service delivery (Du Gay 1997). On the other hand, this individualized organizational involvement produces risks of its own.

To demonstrate the latter I shall make use of a conceptualization of privacy I have developed elsewhere, on the basis of Westin's classical work (Gastelaars 1998; in press; Westin 1970). 'Privacy' is always produced at the interface of institutional surroundings and individual lives. The concept can be associated with the effects of (involuntary) interactions between autonomous individuals and society at large, or between 'private' citizens and public interventions, but it also highlights the vulnerabilities of individual involvement in organizations (Etzioni 1999; Westin 1970). Moreover, I am particularly focusing, here, not only on informational privacy (and IT-supported data-harvesting) as such, but also on the many other, non-virtual, privacy effects produced by everyday negotiations, for instance between representatives of organizations and their clients (Gastelaars 1998; in press). In my opinion, such everyday negotiations render visible the critical edge of individualization in organizational life.

I am particularly interested in the privacy-effects of three organizational repertoires related to the current culturing of service delivery. The first two concern an intensification of the organization's boundary-setting activities: they are (1) the redefinition of the organizations' recruitment and marketing efforts, and (2) an increased interest in selection and classification. The third repertoire concerns the effects of culturing upon (3) internal organizational regimes. In the following section I shall give a first impression of what is meant by these three repertoires. In the next three sections I shall demonstrate empirically how they currently work out.

ORGANIZATIONAL BOUNDARY SETTING AND ORGANIZATIONAL REGIMES

'Organizations' usually try to separate their specific domain from non-organized everyday life; they produce boundary roles in order to accomplish some internal consistency, and also claim to perform 'conscious disciplinary processes whereby organizational leaders adapt internal structures to their perception of external conditions' (Morgan, quoted by Gellner and Hirsch 2001, p. 3). The starting point of this contribution is provided by the observation that these activities are currently becoming more explicit. This particularly is the case in the realm of the human service organizations delivering 'services' on behalf of the welfare state (Gastelaars 1997; Hasenfeld 1983). Three organizational repertoires particularly stand out.

First, and as an explicit reaction to the culturing of service delivery, many organizations will consider an intensification of their *recruiting and marketing efforts*. They redefine their boundary roles, as a part of their efforts to selectively 'target' new employees and clients. They want to get acquainted with them before they take them on. Accordingly, supermarkets are trying to find out 'everything' about our buying behaviour. Nor is it unusual, nowadays, to monitor 'troubled teenagers' extensively for some time, in order to identify them as a 'prime target' for the preventive practices provided by the local police or by other institutions. Such permanent monitoring of prospective participants, clients or otherwise, may first of all affect the so-called *private sphere* that is, 'that sphere of action in which society, as distinguished from the individual, has, if any, only an indirect interest' (Mill [1859] 1975, p. 17) But it may also affect the state of privacy that has been called *anonymity*. Anonymity provides a state in which 'an individual is in public places or performing public acts but still seeks, and finds, freedom from identification and surveillance' (Westin 1970, p. 31). The current debate on informational privacy applies in this context as well.

Second, many organizations, particularly organizations performing service delivery, demonstrate an increasing interest in the *selection* and *classification efforts* they perform, affecting both their clients and their employees. First of all, they want to be more specific as to the 'yes' or 'no' they provide at the organization's door; but, secondly and more importantly, an unprecedented amount of organizational energy is invested, nowadays, in the organizations' permanent monitoring of its – professional and financial – allocations. They want to improve the everyday decision-making processes they perform with their clients as to who gets what, where, when and how. Employees are confronted, accordingly, with efforts to improve upon the 'transparency' of their everyday negotiations. Clients, in their turn, are not only confronted with the potentially stigmatizing effects of the labels that are applied ('suitable for treatment'), their *self-determination* is affected as well. Moreover, it might be worthwhile in this context, to investigate the state of privacy that Westin has labelled *solitude*. This state of privacy is expected to provide people with 'moments of their own' in order to provide them with sufficient private space, to perform their 'dialogue with the mind or conscience' while taking a well-considered decision (Westin 1970, pp. 31–3). And again, the concept of informational privacy also applies here in so far as vulnerable personal data such as medical records are involved (Etzioni 1999).

Third, most organizations have always been quite sensitive to the performance of their internal *organizational regimes* or – to put it in less repressive terms – of the set of preferential attitudes to be performed not only by their personnel, but also by their clients. However, in a manner similar to the two other repertoires, such expectations currently are becoming much more explicit. For instance, the preferential professional attitudes related to the organization's performance of 'client-centredness' are often reinforced by means of so-called corporate cultures re-arranging life 'within'.

Such organizational regimes are not only to be judged by their explicit purposes, but by the extent to which they produce 'cultural intimacy' as well. This notion of cultural intimacy, developed by Herzfeld, applies to the simulacra of familial social relations many organizations appear to perform, providing their constituents with a sense of belonging (Herzfeld 1997, p. 6). Yet it remains to be seen whether the state of privacy that Westin called *intimacy* applies here, too. This state of privacy provides one with opportunities to be 'oneself' when among others, an expectation that can only be fulfilled on the basis of a mutual unquestioning acceptance 'no matter what'. Current managerial attempts at homogeneous corporate cultures often appear to deny the ambiguities that such intimacy may produce. Yet the clients' experiences may be altogether different, here.

For instance, their participation in the organization's performance may be very short term, indeed. In that case, the client's involvement will not be expected to move beyond a general confidence (or 'trust') that the organization will somehow deliver. Consequently, I shall focus on long-term client participation, here. This appears to move between two extremes with, on the one hand, the total incorporation – and a total denial of cultural intimacy – produced by total institutions and, on the other hand, the reciprocal performance, not of 'intimacy' but of 'trust', based on 'open', 'straightforward' negotiations between equals. According to Gutek (1995), the latter is essential to long-term client relationships.

Each of these three organizational repertoires can be taken as exemplary of a specific type of individual involvement in (or commitment to) a given organization, the central theme of this book. So the intensified recruitment and marketing efforts of contemporary service providers may have a demonstrable impact upon the ways in which potential participants may or may not *identify* with a given organization. The organizations' selection and classification efforts, in their turn, may particularly illustrate a given organization's impact upon a person's *participation*, that is, one's active involvement in the allocation of the organization's repertoires. And, finally, corporate cultures or organizational regimes inadvertently relate to one's physical or psychological *attachment* or even one's *loyalty* towards a given organization, and consequently may affect the integrity of the individual. Table 4.1 summarizes this analysis of organizational repertoires, states of privacy and organizational intrusions upon individual involvement.

Both clients and employees can be subject to these three organizational repertoires. Yet in the empirical examples I provide I shall take the client experience as a starting point. Not only because the client experience can be considered to be the ultimate test of the client-centredness of a given organization's performance, but also because these three organizational repertoires also affect the involvement of the organization's employees, sometimes increasing their discretionary powers, sometimes incorporating them to an unprecedented extent. I shall take my empirical examples from the primary processes of service delivery performed by various human service organizations related to the welfare state.[1]

Table 4.1 Organizational repertoires and organizational intrusions upon individual lives

Organizational repertoires	Relevant states of individuality and privacy	Intrusions upon individual involvement in the organization
Recruiting and marketing efforts	The free and uninhibited expression of one's individuality: anonymity and private sphere	Involuntary identification, affecting one's identification with a given organization
Selection and classification efforts	Individual self-determination as opposed to determination by others: empowerment (or self-determination) and solitude (a personal space to reflect)	Dis-empowerment, affecting one's active participation in a given organization
Organizational regimes	The integrity of the individual: the performance of intimacy or trust	Physical and/or psychological harassment, affecting one's attachment and loyalty towards a given organization

ORGANIZATIONAL CURIOSITY AND THE INVOLUNTARY IDENTIFICATION OF PROSPECTIVE CLIENTS (AND EMPLOYEES)

The historical starting point, here, is the complaints by quite a few commercial organizations that they have lost track of their potential clients not because these clients have become physically untraceable, but because of the untraceable nature of their individual desires and orientations (Gabriel and Lang 1995). Particularly, organizations involved with mass service delivery or with mass production of consumer goods took a lead in this realization. They discovered at some point in time that it is not objective data like age distribution and socio-economic differences they had to be interested in but, rather, an immeasurable variety of possible preferences, desires and aspirations. As culture critic, Dick Hebdige summarized this change more than ten years ago:

> we live in a world and in bodies which are deeply scored by the power relations of race and class, sexuality and gender but we also live – whether or not we know it consciously – in a world of style-setters, innovators, sloanes, preppies, empty nesters (working couples with grown up families), dinkies (dual-income-no-kids), casuals, sensibles, the constrained majority, and today's prime targets, the pre-teens and woofies (well-off-older-folks).
>
> (Hebdige 1992, p. 173)

The explicit intention of contemporary organizations actively to anticipate such permanent changes, and improve their marketing and design, have been labelled the culturing of production and of service delivery by Du Gay and others (1997, pp. 236 ff.). This concept implies that organizations not only appear to consider the exchange value and/or user value of the products and services they provide, but their symbolic value, too (Nixon 1997, p. 185). Prospective clients are expected to not only make use of, but also to identify with the organization's offerings. The organizations' anticipation of their customers' 'individual choices' may affect the individuals' *private sphere* in so far as such marketing efforts interfere with their everyday 'private' negotiations. Another notorious effect, however, is the nearly insatiable organizational curiosity performed by service providers. It is particularly the commercial, IT-supported collection of 'personalized' data that re-animated the 1970s' debate about informational privacy.

Westin has indicated that the state of privacy is particularly affected by this kind of curiosity. Anonymity provides a state in which 'an individual is in public places or performing public acts but still seeks, and finds, freedom from identification and surveillance' (Westin 1970, p. 31). As Bauman argues, however, in the commercial sector such data are usually provided on a voluntary basis: 'The storage of massive quantities of data, amplified with every use of a credit card and virtually each act of purchase, results … in a "superpanopticon", but a Panopticon with a difference: the surveilled, supplying the data for storage, are prime and *willing* factors in the surveillance' (Bauman 1998, p. 50). In tune with the so-called 'stranger-effect', once perceived by Georg Simmel as one of the advantages of modern city life, people voluntarily provide such information not expecting to run the risk of identification. But anonymity is not what it used to be: these stored data can always be 'individualized' so chances are that any individual can be identified at any given moment (Gastelaars 2002; in press).

The current debate about informational privacy is triggered by this state of affairs in particular.

Although we encounter a similar IT-supported organizational curiosity in the human service organizations I am discussing here, the data are usually *not* provided voluntarily. For instance, in their Canadian research Ericson and Haggerty demonstrated that, due to the preventive ambitions of their urban police forces, the notion of police surveillance – which used to denote a rather physical activity, with watchful people moving about – has taken on a quite different meaning, anticipating the selection and classification efforts of organizations: 'What we are doing is proactive patrol ... go out in residential areas ... and just write down anything suspicious' (patrol officer, quoted by Ericson and Haggerty 1997, p. 147). Ericson and Haggerty also noted a specific interaction between the computer and the professional:

> It (= the machine) is right there beside you, and it takes two seconds to type a name and a date of birth and you can query anybody. Plus if you query somebody and get a hit on them you can send a message anywhere in Canada to any other CPIC terminal ... You can go along and query plates as you drive around. You are constantly checking anything suspicious. Vehicles. People. If you stop somebody you can query them right on site. I have a tendency to drive down the street and to query anybody. Just query vehicles as they go along ...
>
> (Interviewee, quoted by Ericson and Haggerty, 1997, pp. 140–41)

Moreover, the more individuals set value on their privacy, and the more they are consequently inclined to be very selective in their disclosure of information, the more such organizations are inclined to recover what is being kept from them (Ericson and Haggerty 1997, p. 117). Attempts of street-wise clients to manipulate their own 'visibility' may inadvertently produce even more institutional curiosity.

Yet, there are also human service organizations that appear to perform this initial recruitment and marketing stage in quite a different way. They produce *mutual* and *face-to-face* information strategies, instead of this unlimited organizational curiosity performed back-stage. They also try to physically anticipate upon the future 'organizational experience' of their clients, producing 'preliminary' performances of the professional repertoires in which their clients will be expected to participate (cf. also Nykamp 2001). Especially, secondary schools in the Netherlands do many things to provide an impression of what 'real life at school' will be like for their prospective pupils, and appear to suppress their organizational curiosity at the same time.

> The school primarily approaches the prospective pupils' parents. Not by direct mail – 'it creates a negative image: "are they so much in need of pupils?"'. The

supermarket and birthday-party gossip is also considered to be far too 'unmanageable' for the school. So are the local newspapers, except, of course, for the publication of straightforward data like the announcement of open days, of registration periods et cetera.

The open days themselves are considered very important, and so are the mini-lessons, when both children and parents can participate. On such occasions, when the school relies on face-to-face contact with its prospective clients and, not to forget, their parents, the school can also afford to distribute leaflets because, however 'corny' the children may find them, they do provide their parents with the feeling of being informed.

Physical aspects such as the building – 'transparent, clean, and modern' – are mentioned by many parents as an attraction of the school, just like the (cycling) distance their children will be confronted with, when going to school on their own. The comparatively small, separate building this school provides for the first 'bridging' year of secondary education makes a favourable impression on many parents, too.

Parents are also interested in the school's atmosphere, in the way people deal with one another, in the 'structured' rather than 'loose' regime this school provides, and in the 'individualizing' efforts it is prepared to perform. 'How does the school deal with bullying? Does it provide computer classes? Are there specific programmes for dyslectic children, for highly gifted children? Can the parents of Islamic girls count on separate sports and gymnastics classes under the guidance of a female teacher?' The school is expected to provide parents with information about its organizational regime.

The children are particularly keen on joining their friends: 'my daughter was only interested in the other kids she saw. Children she knew and older than herself. She said all the time "Oh, look! There is so and so!"'

The school itself has its agenda, too. For instance, this school would like to have more pupils with a pre-university level of performance, but does not get them on a regular basis as such children are sent to the smaller, specialized grammar schools nearby. And, whereas other schools might insist on improving their 'output levels' anyway, this school has chosen to actively market the wide range of choice it provides to prospective pupils, apparently taking the side-effects of a somewhat doubtful average academic performance in its stride.

Accordingly, it is only at the later stage of the selection and classification process that the school appears to be interested in personalized information concerning the individual child, such as the child's tests, the references made by the child's elementary schoolteacher; reports on the child's social and emotional development (How does it behave in class? Is it bullied a lot? How is the situation at home? What are the child's strong and weak points?).

(Secondary school, 2000)

Note that the parents are considered to be very important, here. But also note that, on the one hand, this school appears to be well aware that its process of service delivery and its organizational regime will be a part of the future lives of its prospective pupils, whereas, on the other hand, the school must be concerned about its selection and classification outcomes, too: all Dutch schools are rated yearly in terms of their school results by a nation-wide monitoring system. Yet this school does not appear to rely on the performance of any kind of preliminary 'organizational curiosity'.

Two elements may be relevant to explain this. First, free school choice is a very strongly embedded Dutch adage, that used to be a part of the traditional 'pillarization' along religious lines of the state-supported Dutch school system, but fits in with contemporary concepts of 'free consumer choice' as well. Accordingly, the formal possibilities for Dutch schools to be preventively selective are very limited indeed. Second, and more central to the present argument: some service providers appear to fare much better with a 'face-to-face' approach in their marketing. This may have to do with their clients' voluntary access, but also with their clients' prospects in the long run. This school particularly appears to anticipate upon the future performance of parents and children alike.

ORGANIZATIONAL SELECTION AND CLASSIFICATION: DECISION-MAKING, SELF-DETERMINATION AND SOLITUDE

Selection processes merely appear to be designed to establish at the beginning of the process of service delivery whether or not the organization has something to offer (Gastelaars 1997; Hasenfeld 1983). In practice, however, this question is quite often framed in the reverse: which people should be barred from this organization and who can be let in? Even if people approach the organization on a voluntary basis, in the end they may not be allowed to enter. As Bauman argues once more, particularly in the commercial world this so-called risk selection of potential customers is there to 'confirm their reliability as clients and *choosers*', and to sift out 'those incapable of choice ... before damage is done or resources wasted' (Bauman 1998, p. 51). No wonder that many of us have become quite sensitive to their potentially discriminatory effects.

Yet there is also another organizational activity that appears to be very similar to these selection efforts, but in fact is not. It is organizational classification. Here the relevant questions are: once this individual is let in, what are we going to do with her or him? Or: what is he/she going to do for us? And: are we going to continue or change the course of action taken? In fact, these processes are often far more complicated than the selection process. They concern the specific translation of the cultural complexities of the external world and of the everyday negotiations of prospective clients (and also of prospective employees) into the organization's repertoires. And they concern negotiations around the 'progress' made by a client (or an employee). Moreover, the decision-making processes involved do appear to be particularly one-sided. Due to their complicated nature, they usually are performed by highly qualified professionals and – from a client perspective – they are performed back-stage.

As a matter of fact, such classificatory efforts do not necessarily produce a homogeneous 'we'. On the contrary, they appear to define the range of accepted diversity 'within'. According to the American anthropologist, Herzfeld, such a mechanical management of difference even may create its own sense of belonging (Herzfeld 1992). It appears to invite individuals to 'make a difference' without being excluded.

However, under the threefold pressure of (1) trying to adapt the organization's offerings to individual clients' aspirations or needs ('culturing'), (2) the need to produce more transparency as to the allocations the organization performs ('accountability') and (3) the explicit development of the appropriate professional competencies, most street level bureaucracies and professional service providers related to the welfare state are currently reinforcing their 'classificatory' repertoires to an unprecedented extent.

An ever-increasing amount of organizational energy is spent, for instance, on IT-supported decision making systems performed 'back-stage' (Bovens and Zouridis 2002), apparently overruling the professionals' prominence in the decision-making processes that are involved. The latter is, for instance, very visible in the realm of the street-level 'selection bureaucracies' providing social security.

> The new law on Basic Social Security (*Abw*) of 1996 expects the Social Services (*Sociale diensten*) to not only to perform their original core businesses, which is to establish an individual's entitlement to a modest basic income. The new law also expects these Social Services, and basically for reasons of governmental cost-reduction, to promote the same person's expedient discharge from the institution. This second 'core task', then, is to promote their clients' successful return to the labour market. This latter activity is locally called the 'outflow', the 'participation', the 'activation' or 'reintegration' function of the Social Service.
>
> Accordingly, the organizations are expected to provide 'individual trajectories' through which individual clients are expected to move. An individual client's entitlement to such 'intensive care', however, should be very carefully assessed. That is why such an entitlement is established, among other things, by the application of an IT-supported Opportunity Yardstick (*Kansmeter*) that 'establishes for each and every unemployed, according to a number of criteria such as one's level of education, working experience, social empowerment, and one's physical and psychological disabilities' the so-called 'distance from the labour market' or 'Phase'.
>
> People in Phase I are expected to be able to fend for themselves; Phase IV unemployed are considered to be excessively problematic and may be relegated to

> specialized institutions. However, for unemployed people allotted to Phase II and III a 'uniformly qualifying intake' is performed – the so-called Quint (or *Kwint*, in Dutch) – in order to refine the insight regarding the individual's prospects on the labour market. The outcomes of these procedures determine the organizations' offer of training, help and care. It is only after the performance of these extensive computerized negotiations, that more elaborate 'face to face' negotiations are provided to individual unemployed. Moreover, the professionals appointed for this work are explicitly admonished 'not to negotiate for too long'.
>
> (Social Services 2001)

What at first sight appears to be an effort to perform 'individualized' attention, works out as a computerized selection and classification effort, in which the clients' role appears to be restricted to that of providing the appropriate data, for the extensive and individualized information systems that are employed back-stage. Moreover, such IT-supported classificatory infrastructures may overtly be designed to make the decision making processes more transparent, yet they tend to 'fade into the woodwork' from a client perspective. Hence, as Rifkin observed for the commercial world, 'the customer … may become totally dependent on … forces that he or she little understands and over which he or she has less and less control' (Rifkin 2000, p. 103). To Rifkin, the latter circumstance explains the current popularity of such *personae* as the 'case manager', 'agent' or 'coach' that can be observed in many human service organizations as well.

And, although one might expect in this era of individualization, that these gatekeepers allow their clients to exercize their 'freedom of choice' and their 'right of self-determination', in the end 'the client … may become increasingly dependent on the "expert" agents over time to manage his or her affairs' (Rifkin 2000, p. 104). Obviously, the clients' participation, or self-determination, is at stake.

In our law-orientated society, we tend to put our hopes on the law, in order to empower individuals to negotiate the undesirable effects of such organizational classifications. Under the present informational privacy laws, for instance, individuals are entitled to inspect their own files and to retrieve the information collected about them. People also have acquired the right to refuse disclosure of 'personal information', pertaining to 'those facts, communications, or opinions which relate to the individual and which would be reasonable to expect him to regard as intimate or sensitive and therefore to want to withhold or at least to restrict their collection, use or circulation' (Wacks [1989] 1993, p. 26; cf. also Etzioni 1999). Moreover, individuals are now also entitled to re-negotiate selection and classification processes they deem hurtful, by means of formalized 'informed consent', 'second opinions' and 'complaints'. In the end, however, such efforts may not mitigate the effects of the power-dependencies produced by such classification processes.

More helpful in this context, then, and certainly more in tune with the present focus on everyday negotiations, is the state of privacy Westin called the individual's entitlement to solitude. This is 'the voluntary and temporary withdrawal of a person from society through physical or psychological means' (Westin 1970, p. 7) providing a person with sufficient space to reflect, to perform one's own preferences and to 'avoid being manipulated or dominated wholly by others' (Westin 1970, p. 33) when coming to terms with a decision.

But even this does not turn such classification processes into *mutual* (and possibly even *reciprocal*) affairs in which both negotiating partners appear to have an equal say. Such a preoccupation with the relational quality of service delivery is not uncommon in the realm of commercial service delivery. Barbara Gutek discusses, for instance, its relevance for long-term relationships, for instance, between high-ranking solicitors and their corporate clients. The mutual commitment of both customer and provider is even considered essential and should be established from the start (Gutek 1995, p. 17). Then the power-dependencies are expected to be less explicit, or maybe even reversed in favour of the client.

But it is much more difficult to come up with similar examples, of such 'explicit', 'open', 'straightforward' and 'reciprocal' negotiations between two equal parties, in the sphere of the human service organizations I am discussing here. For instance, as I have demonstrated in the previous paragraph, however 'free' the school choice in the Netherlands appears to be, once a child is admitted to the school and the educational classification processes – tests, examinations, monitoring – can start, the children (and their parents) are subject to the institution's considerations. Even the face-to-face negotiations human service organizations provide usually are negotiations-in-dependency, in which the expert power of the organization's representatives usually plays a decisive role. Moreover, the explicit performance of solitude – providing for some space (and time) to help people make up their own balance of pros and cons – requires an extra effort by the professionals.

> Even long before IT-supported pupil-monitoring systems were developed, 'Each achievement an individual pupil performs in class is evaluated, recorded and expected to have its future consequences in terms of the pupil's intellectual career'. The school performs its classifications back-stage. However, there are many individual choices to be made and they may become subject to more or less open negotiations. The children and their parents are confronted with a large number of options as far as the choice of subjects is concerned, and they have to be matched, not only with the child's individual preferences but also with its capacities and its level of performance, as judged by the school. Such classificatory negotiations require new professional competencies apparently

> designed to maintain the organization's lead under all circumstances: 'New teachers take a training course in "parent-teacher encounters": with do's and don'ts for the individual parent-teacher hours. These do's and don'ts are sort of standardized nowadays, because of the parents' and their children's empowerment.'
>
> (Secondary school 2000)

Although parents appear to be truly respected as the 'natural' coaches of their children, in this classificatory context the power-dependencies between the organization's representatives and their clients appear to dominate everything else.

THE ORGANIZATIONAL REGIME AND THE PRODUCTION OF INTIMACY AND TRUST

People who work in an organization, but also people who participate as clients, are always subject to an organizational regime. Such organizational regimes usually perform a large number of 'tacit' organizational routines but they also perform the organization's explicit rules and regulations. A lot of studies of organizational culture deal with precisely this subject matter (Martin 2002). However, there currently are many managerial attempts that claim – often on behalf of client-centredness, and of the associated need to 're-culture' their organizations – to transform the most implicit or 'tacit' of their routines.

The students of organizational culture have been very quick to point out, of course, that such 'homogenizing' efforts tend to ignore (and even negate) the complexity, multi-layeredness and, above all, multi-perspectivity that underlies the performance of culture (cf. De Ruijter 2000; Koot 1995; Verweel 2000). Verweel even speaks of the 'mechanization of the social', and points out that the instrumental (or 'rational') and the negotiation (or 'social') aspects of organizational life may become dangerously intertwined (Verweel 2000, p. 9 ff.). And, indeed, the performance of corporate culture usually combines a top-down prescription of a preferential culture (that is, attitudinal 'rules and regulations') with an equally top-down performance of an organizational 'we'. The latter is expected to enhance each individual's involvement in the organization (cf. Deal and Kennedy 1999). And although the American anthropologist Herzfeld still maintains that even such 'discriminatory' processes may provide the organization's constituents with a logic that 'naturally' confirms who belongs and who does not (Herzfeld 1997, p. 7) the ways people actually deal with such propositions may be very different, indeed.

The state of privacy that Westin calls *intimacy* may provide us with a critical point of reference, here. This state of privacy provides an individual, within a given community, with opportunities to be 'him- or herself', *no matter what.* It is, however, precisely this 'no matter what', this mutual *unquestioned* acceptance, that produces the kind of 'cultural intimacy' Herzfeld had in mind when discussing the individual's involvement in a given organization. According to him, such cultural intimacy only works out to the extent that it provides the 'insiders' with sufficient opportunities to deal with the ambiguities that inevitably come with belonging (Herzfeld 1997, p. 9). Accordingly it may create opportunities to allow 'one of us' to temporarily fall out, or even to reframe someone's 'bad' behaviour within an acceptable range of mixed feelings: 'Of course he is a liar and a cheat, but he is our President, you know!' Concepts like corporate culture tend to ignore such ambiguities, and consequently may not be effective at all.

From a client perspective, however, the situation may be different. First of all, as Gutek (1995) has pointed out, a client relationship may have very little impact and may be of very short duration. And, although 'trusted' relationships are often mimicked in such short-term and superficial organizational transactions Gutek calls encounters, they should not be taken for the real thing. Even if a call centre employee succeeds in enticing her client with all kinds of personalized details, the client himself should be well-advised to realize that no personal intentions are involved, at least not on behalf of the organization's representative: none of this would happen if his number failed to show up.

Yet, and secondly, with the kind of long-term relationships I am focusing on, the situation may be quite different, indeed. The most extreme examples of an organizational regime, from a perspective of clients' vulnerabilities, are obviously provided by the so-called total institutions providing care for 24 hours, seven days a week (Goffman 1961). The physical superiority, the all-encompassing relationships and the tendency to maintain one-sided communication patterns that Goffman attributed to such institutions, tend to ignore all aspects of 'cultural intimacy', and consequently may fail to produce the involvement of their clients. We take an example, here, with the client perspective as a starting point, from a research report comparing a number of asylum seekers' residences in the Netherlands.

> Two of the locations consist of 'concentrated' residences, that is, traditional asylum seekers' centres where a large number of people and activities are concentrated in one location.

There, the asylum seekers appear to be very interested, indeed, in the physical aspects of privacy, that enable them to physically (or psychologically) withdraw from the presence of others. 'I have a room of my own, so I do enjoy a bit of privacy.' Particularly in such overcrowded settings as a former Catholic convent, the proximity of many others is experienced as quite oppressive. There are many complaints about disturbances by others, and many apparently irrelevant fights. Interestingly, the employees themselves appear to seek privacy from the ever-present asylum seekers, too, avoiding them for instance by means of locked doors and strict office hours.

Moreover, in these locations the theme of privacy also appears to relate to a lack of anonymity and to a sense of permanent surveillance. The asylum seekers themselves can't avoid the impression of permanently being watched. And rightly so, as the organizational regime tends to punish and reward individual behaviour. It uses such classificatory means as the 'withdrawal' or the 'granting' of specific allowances, or 'transfers' to better locations or to locations that are even worse, as a means to effectuate the compliance of their clients. In fact, most asylum seekers even appear to think that their behaviour may directly affect their legal procedure, although formally this is not the case. Apparently, in the everyday negotiations, such classificatory issues prevail.

(Asylum seekers' residences 2001)

The organizational regime appears to focus on surveillance, and no cultural intimacy is to be found. Moreover, it would appear to be far-fetched to apply – to the mutual relations of the asylum seekers but also to the relations between asylum seekers and the responsible employees on location – the state of privacy Westin called reserve, that is, the 'willing discretion of others' protecting the individual's need to limit communication about himself (Westin 1970, p. 32): 'indifference' or 'mutual avoidance' are far better terms, in this case. And, finally, particularly when we consider the negotiations between asylum seekers and the responsible professionals on location, the power-dependencies are decisive: asylum seekers feel, and are, delivered to the organization's classificatory regime.

In two other locations presented in the same report the asylum seekers are allowed to live in independent private houses in regular neighbourhoods.

The self-chosen intimacy appears to be relevant here, and so does the private sphere. The asylum seekers clearly appreciate living on their own as a family. 'It suffices that we can be on our own as a family. In the caravan site (where we used to stay) every three or four months some new people arrived, to share our caravan. We hardly managed. Thank God we are together, now.'

The asylum seekers also appreciate the fact that they are much less confronted on a face-to-face basis with the professionals responsible for their legal procedures and for their everyday residential care. They now have to register at a distant Aliens Office once a month instead of every week. The maintenance men working for the organization providing the residences are visiting them on a

> weekly basis. The 'sanctioning powers' remain with the 'distant' professionals responsible for their residential care.
>
> The latter develop new repertoires, accordingly. They are learning to let the asylum seekers fend for themselves, instead of always helping them. Moreover, they appreciate this opportunity to actively 'distance' themselves from the asylum seekers. The new physical distance helps them for instance to avoid emotional involvement. The fact that they are still endowed with their sanctioning powers is less predominant, in their experience.
>
> Yet, although the asylums seekers appear to appreciate their relative freedom, they still experience the weekly visits of the maintenance men in their private houses as a face-to-face surveillance. Moreover, the 'sanctioning powers' of the professionals are felt to be present, too.
>
> (Asylum seekers' residences 2001)

This aspect of power-dependency may be less dominant in organizations where clients enter on a more voluntary basis, and are also expected to actively participate in the process of service delivery itself. Then Gutek's notion of a 'relationship' may be approximated, to an extent, and one can expect these negotiations to be much more on an equal footing. An example is once more provided by the negotiations between parents and their children and between parents and the representatives of our secondary school, concerning the (individual) interpretation and application of the school's disciplinary rules.

> One parent particularly appears to be very preoccupied with the empowerment of her child, even considers this an issue for the school curriculum. About a conflict between her child and a teacher: 'Of course, I am there when my child says, "I can't do it on my own"; then I say, "I don't mind preparing together for this conversation, but as I see it, you eventually have to do it on your own; I'll only come with you if it doesn't work out". And, about the school's attendance rules: 'I think the children themselves are responsible … they … have to learn, "what do I need, where am I to go, what had I better do on my own" … you have to be able to make mistakes.' She adds: 'our educational system is focused too much on the cognitive, the specialized subjects and too little on … social competencies, being respectful with each other, taking responsibilities'.
>
> Another parent appears to be less self-assured. Her child has good grades but doesn't appear to behave. 'I tell him, he shouldn't protest in class but do it after hours!' She finds the parent-teacher hours rather stressful 'when they talk, it is not about his grades but about his attitude and behaviour'. Yet she insists that 'they inform us, I have to take care not to start talking myself and tell them … it is good to also get someone else's impressions of K. … and there are many points we recognize'. But she also appears to feel quite powerless: 'It's so typical, I think, for teachers to say "But his grades are OK, aren't they?" There is more to it than grades, that's what I say.'
>
> (Secondary school 2000)

Once more, the parents are respected as the natural coaches of their children. Moreover, they are helped by the fact that the school apparently is 'cultured' in the sense that contemporary 'democratic' familial relations are also reproduced within the school, so that both parents and children can be expected to rely on the school's performance in this respect. The school apparently provides a strongly 'negotiating' regime.

The quotes also illustrate that teachers, in their face-to-face negotiations with individual parents, are expected to switch roles all the time, to make room for an individual variety of preferences and expectations, and yet 'manage' the complicated interactions that ensue. In such everyday negotiations the organization's everyday performance with its clients (or their representatives) apparently requires much more than 'leaving them alone'. It even appears to require much more than a generalized acceptance 'no matter what': it requires an active and reciprocal negotiation of individual preferences and performance, an overt and non-judgemental dealing with diversity 'on the spot'.

CONCLUSION

In this contribution I have discussed three organizational repertoires. In fact, these repertoires – relating to the organization's boundary roles and to the organizational regime – also stand for the client's movement *through* the primary process of service delivery, beginning with the recruitment phase and, next, working through the organization's selection and classification efforts, and, finally, arriving at the organizational regime performing the intended 'transformation' of the client (cf. Gastelaars 1997; Hasenfeld 1983). Accordingly, there will be very few clients that will not be confronted, somehow, with some organizational restrictions. Their involvement in the organization will always be under some duress.

Yet, on the level of the primary processes of service delivery, the organization's responses to the 'culturing' of service delivery can be very different indeed. Depending on the organization (and its core business), organizations may be particularly inclined to provide more or less room for individualized negotiations, and for the performance of 'individuality'. In fact, the organizational response to culturing may move in two opposite directions, in this respect.

The starting point for the first kind of response is provided by the classificatory efforts organizations currently perform. If 'culturing' means that organizations are to be expected to somehow negotiate the very diverse individualities of their clients, then 'classification' should be seen as the ultimate organizational repertoire to keep such diversity under control. From a client perspective, such classificatory efforts often tend to move 'back-stage'. And even when they are performed in face-to-face negotiations, they yet appear to confirm the organization's dominance over its clients. Sometimes they are performed on an equal footing, but more often the power-dependencies prevail. And although sometimes they even may develop into mutual or even reciprocal negotiations, more often they are performed in a one-sided manner or even top down. Moreover, whenever these classificatory issues prevail in a given organization – as I have demonstrated with the Social Security providers and the asylum seekers' residences, and also with the metropolitan police, although a situation like that may very well emerge in professional organizations like hospitals and schools, as well – then it may also be expected that even the recruitment efforts these organizations perform, do anticipate their classificatory efforts. It also may be expected – as I particularly demonstrated in the 'concentrated' asylum seekers' residences – that such classificatory issues may pervade the everyday negotiations performing the organizational regime. A classificatory regime 'mechanically' provides for individualization, and tends to restrain the performance of individuality in many ways.

Yet, depending on the situation, there is also another organizational response to the 'culturing' of service delivery that stands out. I have demonstrated – particularly with the secondary school, but again, in other organizations such a regime may prevail as well – that in organizations where both the voluntary commitment of their clients and their clients' active participation in the everyday negotiations are considered a condition *sine qua non*, the situation may turn out to be quite different. In cases like that it can be expected that the organization's recruiting efforts are particularly affected not by classificatory efforts, but by the anticipation upon the clients' *experience*, including one's future participation in the organization's everyday regime. Of course, such organizations are expected to classify as well, but they may do so in an entirely different manner. For instance, they may be prepared to provide for mutual information strategies, and mutual negotiations, when performing their classifications. And if they do not effectively empower their clients, then at least they may provide for some solitude, to properly negotiate the various decisions. And even if their regime does not provide for intimacy (or trust), then at least it may provide some room to negotiate diversity on an individual basis.

If 'culturing' means that organizations are to be expected to actively negotiate the very diverse individualities of their clients, in cases like this the organization appears to choose for active and explicitly 'individualized' negotiations, instead of classificatory control.

NOTE

1. The sources providing the empirical illustrations are listed with the references.

REFERENCES

Bauman, Z. (1998), *Globalization. The Human Consequences*, Cambridge: Polity Press.

Bovens, M. A. P. and S. Zouridis (2002), 'Van street-level bureaucratie naar systeem-level bureaucracy. Over ICT, ambtelijke discretie en de democratische rechtsstaat', *NJb*, **2**, 11 January, 65–74.

De Ruijter, A. (2000), *De multiculturele arena*, Tilburg: Katholieke Universiteit Brabant.

Deal, T. and A. Kennedy (1999), *The New Corporate Cultures: Revitalising the Workplace after Downsizing, Mergers and Reengineering*, London and New York: Texere.

Du Gay, P. (ed.) (1997), *Production of Culture/Cultures of Production*, London, Thousand Oaks, New Delhi: Sage.

Ericson, R. V. and K. D. Haggerty (1997), *Policing the Risk Society*, Oxford: Clarendon Press.

Etzioni, A. (1999), *The Limits of Privacy*, New York: Basic Books.

Gabriel, Y. and T. Lang (1995), *The Unmanageable Consumer. Contemporary Consumption and its Fragmentations*, London, Thousand Oaks, New Delhi: Sage.

Gastelaars, M. (1997), *'Human service' in veelvoud. Een typologie van dienstverlenende organisaties*, Utrecht: SWP.

Gastelaars, M. (1998), 'Privacy gelokaliseerd. Over systeemdwang, zelfbeschikkingsrecht en goede manieren', in Rathenau Instituut (ed.) *Privacy geregistreerd*, Amsterdam: Otto Cramwinckel, pp. 155–87.

Gastelaars, M. (2002), 'How do statistical aggregates work? About the local effects of general classifications', in Barbara Czarniawska and Heather Höpfl (eds), *Casting the Other: The Production and Maintenance of Inequality in Work Organizations*, London: Routledge, pp. 10–33.

Gastelaars, M. (in press), 'About individual autonomy and the relational self: the performance of privacy in contemporary organizations', in Ad van Iterson, Willem Mastenbroek, Tim Newton and Dennis Smith (eds), *The*

Civilized Organization: Norbert Elias and the Future of Organization Studies, Amsterdam: Benjamins Publishing Company.

Gellner, D. N. and E. Hirsch (eds) (2001), *Inside Organizations: Anthropologists at Work*, Oxford and New York: Berg.

Goffman, E. (1961), *Asylums*, New York, London, Toronto, Sydney, Auckland: Anchor Press Doubleday.

Gutek, B. (1995), *The Dynamics of Service. Reflections on the Nature of Customer/Provider Interactions*, San Francisco: Jossey-Bass.

Hasenfeld, Y. (1983), *Human Service Organizations*, Englewood Cliffs, US: Prentice Hall.

Hebdige, D. (1992), 'After the masses', in David Mercer (ed.) *Managing the External Environment: A Strategic Perspective*, London, Newbury Park, New Delhi: Sage, pp. 166–74.

Herzfeld, M. (1992), *The Social Production of Indifference: Exploring the Symbolic Roots of Western Bureaucracy*, Chicago and London: University of Chicago Press.

Herzfeld, M. (1997), *Cultural Intimacy: Social Poetics in the Nation State*, New York and London: Routledge.

Koot, W. C. J. (1995), *De complexiteit van het alledaagse. Een anthropologisch perspectief op organisaties*, Bussom: Coutinho.

Martin, J. (2002), *Organizational Culture: Mapping the Terrain*, Thousand Oaks, London, New Delhi: Sage.

Mill, J. S. [1859] (1975), 'On liberty', in John Stuart Mill, *Three Essays*, London, Oxford, New York: Oxford University Press, pp. 5–141.

Nixon, S. (1997), 'Circulating nature', in Paul du Gay (ed.), *Production of Culture/Cultures of Production*, London, Thousand Oaks, New Delhi: Sage, pp. 179–236.

Nykamp, M. (2001), *The Customer Differential: The Complete Guide to Implementing Customer Relationship Management*, New York: Amacom.

Rifkin, J. (2000), *The Age of Access: How the Shift from Ownership to Access is Transforming Capitalism*, London, New York: Penguin Books.

Verweel, P. (2000), *Betekenisgeving in organisatiestudies. De mechanisering van het sociale*, Utrecht: ISOR.

Wacks, R. [1989] (1993), *Personal Information: Privacy and the Law*, revised paperback edition, Oxford: Clarendon Press.

Westin, A. F. (1970), *Privacy and Freedom*, New York: Atheneum.

Empirical reports

De Kort, I., *Organisatie: Eckartcollege*, Universiteit Utrecht, 2000 (Secondary School 2000).

Du Pree – van der Slik, A., *Casemanagement bij de sociale dienst*, Universiteit Utrecht, 2001 (Social Services 2001).

Gastelaars, M., K. Geuijen, J. van der Horst en M. van Leeuwen, *Tussen arena en netwerk*, Voorlopig eindrapport Onderzoeksproject Kwaliteit Kleinschalige Opvang, USG/ COA mei 2001 (Asylum seekers' residences 2001).

5. Ideology, Consensus and Organizational Effectiveness in a Fragmenting Society: Western Combatants in the Second World War

Keith Grint

INTRODUCTION

The premise of the problems of organizational binding and identity is that the postmodern world has created an array of issues that have thrown our traditional world into turmoil. Two particular issues appear to be critical:

1. *Detraditionalization* The individualization and detraditionalization that has emancipated many has also destabilized the contemporary world and generated an increase in multiple and changeable identities. In effect, the loss of tradition has left people searching for new identities and into this void have stepped several actors and institutions keen to provide any such 'lost souls' with a 'lifeline to normality'. These lifelines comprise the kind of 'strong cultures' beloved of Peters and Waterman (1982), or those provided by the resurgence of nationalism or religious fundamentalism, or even those derived from adherence to any of the variety of forms of New Age spiritualism. This simultaneous fragmentation and hybridization is most evident in areas like the former Yugoslavia which has fragmented but the elements have formed radically xenophobic communities.
2. *Globalization* The globalization of the world economy, and its related regionalization, has also played its part in unsettling what counted as 'normality' and in stimulating the search for stability. Technological developments, such as the Internet and telecommunications, have supported, sustained and encouraged the detraditionalization and globalization, and mixed the virtual with the unreal.

The first issue to consider, then, is whether the current situation is novel because there is a tendency in each generation to see 'revolutions' that hindsight pronounces as merely marginal change. Certainly there have been many periods in the past where traditions were uprooted and whole societies either destabilized or even destroyed. For all the contemporary concerns about the current morass of instability, two issues are worth considering before we fall in line. First, it is difficult to examine any period of history without coming to the conclusion that instability not stability is the norm. That instability may manifest itself in terms of the wars that have plagued Europe, for example, throughout the entirety of the historical record, or in terms of the diseases that have devastated whole populations, or simply in terms of the cycles of famine and other natural disasters. Ironically, therefore, one would have to consider the current era as one of the most, not the least, stable on these criteria. Indeed, one would be hard pressed to argue that the period 1939–45 was *more* stable than the current era. Nonetheless, the perception of instability certainly increased as we approached the new millennium and there can be little doubt that the September 11 attacks and the subsequent events in Afghanistan and elsewhere accelerated the perception of instability – and that perception may be far more important than any concern about the 'reality' of material change. If this is the case, then the relationship between instability and ideological commitment can also best be examined by considering how those directly involved in the Second World War coped with the problems that allegedly plague our own times.

The second issue is globalization. There are clearly higher levels of global integration sustained by internet technologies, financial exchange and world trade but if we take globalization to mean the proportion of gross domestic product (GDP) taken by exports, labour mobility across national boundaries (for example, mass emigration and immigration), and convergent commodity prices, then the late nineteenth century was more globalized than is today's economy (Rodrik 1997). Of course, the level of technological infrastructure has never been as great as it is today, but the revolutionary significance of the internet should be evaluated against the equally revolutionary infrastructure provided before, for example, the railways – heralded as the annihilation of space and time – and the telegraph system – the highway of thought – in the mid to late nineteenth century (Standage 1999). These forms of technologically determinist models of change remain unable to explain, let alone predict, change, and we do not need to situate technology as the causal nexus of globalization to see that other eras have been similarly drawn into one apparent world unit. For instance, the two world wars, and especially the second, undoubtedly generated enormous pressures for global integration and disintegration simultaneously.

But the issue here is not really whether we face a unique situation but, rather, what does a novel situation, or at least the perception of a novel situation, mean for forms of organizational identity or binding? To what extent does it really matter whether the situation is unique or revolutionary or global? Indeed, what counts as unique or revolutionary or 'normal' or 'global' lies in the hands of those with sufficient power to determine what criteria will be adopted and to pronounce on that criteria as such, rather than within the 'objective' situation itself. In short, there is no mechanical or deterministic link between 'the situation' and the forms of identity because identity, as Anderson (1983) has insisted, lies in the imagination not in the so-called 'real world'. Since we can never really 'know' our neighbours or compatriots or fellow citizens to any degree of intimacy, we have to imagine that we have something in common, some bond that unites us. But that bonding is almost always easier to construct and maintain when it is developed *against* 'the other'. Or, as Rowe (2000, p. 324) suggests, 'The greater the threat to our meaning structure the greater the likelihood of focusing on enemies and not friends'. Thus, for example, Colley (1994) and Pears (1997) have shown how 'British' (as opposed to 'English') identity was developed in the context of a war against the French.

The consequence of this is to suggest that the concern for the way 'normality' – that is, stability – generates identity and commitment may be misplaced. In other words, if identity and commitment derive from and are reinforced by stability, then we should expect that a stable society rooted in slavery would generate strong commitments towards particular identities. But how would this be manifest? Would we expect Roman, Greek and African-American slaves to construct either strongly supportive or resistant identities to their stable situation, because there is little evidence that they did. What we can suggest is that short periods of political instability do appear to generate very powerful – and very short-lived – forms of commitment wherein slaves are willing to risk their lives for the sake of freedom, but this is not a permanent state of consciousness. Indeed, Rosa Luxemburg's concept of revolutionary consciousness supports this, for periods of ideological cohesion that generate political revolutions can emerge with extraordinary speed and dissipate just as rapidly. And it is this revolutionary consciousness which destabilizes normality and not the absence of normality that destabilizes and radicalizes consciousness. In effect, if identity and commitment are not a function of the stability or instability of the environment but, rather, act independently of this and, in fact, act upon that environment, then a concern for the ideological consequences of environmental instability is both deterministically mechanical and misplaced. Identity and commitment, in other words, are not mechanically derived from the environment and therefore created and discovered in it, but are imposed upon it.

Let us, however, assume that the situation is one of great flux and novelty. *If* this is the case what does it imply for identity and organizational cohesion? That is, what kind of leadership and organizational control structures work best in situations of uncertainty? Perhaps a prerequisite question would be whether ideological consensus is necessary for organizational efficiency in any situation, and if so when? There is, of course, a moral issue to be resolved here for it may well be that an ideological consensus can be achieved and that organizational efficiency is a result, but to what purpose? Probably one of the most immoral examples of this couplet is the infamous Wannsee Conference, 20 January 1942, in which the Nazi hierarchy demonstrated a remarkable degree of ideological consensus and organizational efficiency to agree upon the foundations for the elimination of Jews (Roseman 2002).

Beyond the ethical issue there is a practical problem of utility: to what extent is ideological cohesion a function of what occurs within an organization? For example, Goldthorpe et al. (1968; 1969) and Fox (1971; 1974) insisted many years ago that the ideologies, or 'orientations to work' or 'frames of reference', that were manifest at work were themselves as much a product of traditions and cultures that existed outside the factory gates. In fact, one need only consider the limited impact of the mass surveys undertaken in the 1930s in the Hawthorne Experiments to remember that it was already apparent to organizational researchers that the non-work environment was critical to performance at work (Grint 1998, pp. 121–2). Of course, it may even be that ideological cohesion serves an altogether different function from the one associated with it in most management texts. That is to say, as Abercrombie, Hill and Turner (1980) argued 20 years ago, such a dominant ideology serves to protect the anxieties of the management far more than it persuades subordinates and employees that the ideology represents both the truth and their best interests.

Alternatively, one may consider that an ideological consensus is actually unnecessary for organizational effectiveness. F. W. Taylor implied that such a 'mental revolution' was essential for achieving maximum efficiency but he recognized that simply fulfilling the material interests of employees would be sufficient to increase efficiency well beyond the limits he found at the turn of the last century. But the limits and disadvantages of ideological cohesion and organizational 'binding' should not dissuade us from examining some uncomfortable issues:

In what may be considered extreme circumstances (war, business crisis, general strike and so on), ideological cohesion appears to provide a significant advantage to the side that can best reproduce it. The ideological cohesion is usually associated with the construction of a hostile, inferior, evil and dangerous 'other', to which many problems and events are attributed on little or no evidence.

That cohesion remains a double-edged weapon. It is often associated with 'group think' (Janis 1982) and can quickly undermine the advantages that it originally provided. As Milgram's experiments suggested, a relatively small proportion of individuals are willing and able to resist manifestations of authority, especially when one's career or life is at stake, and only the existence of a collective ability to resist coercion, what Milgram (1974, p. 118) calls 'Defiant Peers', seems capable of deterring leaders from making catastrophic mistakes.

The paradox of critical supporters, or constructive dissenters, remains fundamental to leaders intent on maximizing their organizational effectiveness: how can you build an ideologically coherent organization that remains composed of critical supporters, that is followers who are both supporters and yet independent enough to remain constructively critical of their leaders?

Such discomfort is probably best examined by adopting Goldthorpe et al.'s (1968; 1969) rationale for analysing the most extreme case of the hypothesis being examined. In their case they were trying to establish the extent to which working-class men adopted middle-class attitudes as their level of affluence increased, in line with Zweig's original *embourgeoisement* thesis. Thus they chose car workers from the Luton Vauxhall plant, then amongst the highest paid manual workers in Britain. In what follows I will take an equivalent approach to consider the extent to which the Second World War, and in particular D-Day and the invasion of Normandy, provides a test bed for the significance of ideological cohesion to organizational effectiveness in times of apparent radical change.

HIERARCHY, HETERARCHY AND ORGANIZATIONAL EFFECTIVENESS

We can begin by assessing the difficulty of measuring the extent to which leadership or ideological cohesion makes any difference because there are so many variables involved and because the measurement systems are as much concerned with constructing the world as with measuring what they construct. This is even more difficult when those involved in the situation are more concerned with staying alive than noting down evidence about leadership or ideological cohesion and so on. But we can start with an argument about the effectiveness of the two opposing armies. Colonel Trevor Dupuy, an American, suggested in 1977 that:

> On a man for man basis, the German ground soldier consistently inflicted casualties at about a 50 per cent higher rate than they incurred from the opposing British and American troops UNDER ALL CIRCUMSTANCES (emphasis in original). This was true when they were attacking and when they were defending, when they had a local numerical superiority and when, as was usually the case, they were outnumbered, when they had air superiority, and when they did not, when they won and when they lost.
>
> (Grossman 1996, pp. 161–2)

It is also worth considering the historical context for this: in the First World War the Central Powers inflicted at the very least 10 per cent more casualties upon the Allies than vice versa, and when the comparison is made between German and Allied troops the disproportionate success of the former increases. For example, Dupuy suggests that German troops were 20 per cent more effective than either the British or the American troops (Dupuy 1977, pp. 328–32). The implication is that even before the SS and the Nazi ideology took hold, the German troops were more effective than the Allied troops – and had already demonstrated this in the previous war.

Hastings (1993, p. 370) concurs, calling the German army:

> The finest fighting army of the war, one of the greatest the world has ever seen. This is a simple truth that some soldiers and writers have been reluctant to acknowledge, partly for reasons of national pride, partly because it is a painful concession when the *Wehrmacht* and SS were fighting for one of the most obnoxious regimes of all times.

The Germans certainly did not root their military superiority in a more effective economy or mobilization; on the contrary, their military superiority was achieved against an inferior economy and mobilization. Yet for the most part the German economy, supported by its conquests, survived well during the war. Indeed, food supplies remained stable – thanks to the wholesale transfer of supplies from occupied countries, particularly the Netherlands (Badsey 1994a, p. 283). Hence while the combat efficiency of the German troops was usually higher than their enemies, the strategic effectiveness and economic mobilization was often palpably worse. There are several reasons generated for the assessment of German military superiority.

First, there is an argument that ideological cohesion prospers in, and negates the need for, a formal and inflexible hierarchy. For instance, the proportion of officers to soldiers was markedly different between the main protagonists in the west. The German army had a lower proportion of officers to soldiers and this necessarily put more emphasis on NCO leadership. For example, the US 29th Infantry Division had 14 281 members, of whom 747 were officers (5.2 per cent). Of the 13 027 members of the German 352nd Infantry Division, only 405 were officers (3.1 per cent). In general, the British Army was even more top heavy and operated with officers making up 6.6 per cent of the total at the beginning of the war, climbing to as many as 7.7 per cent of its total numbers by 1945 (Fraser 1999, p. 100). In other words, the German army had effectively *deeper* leadership than either of the main Allies: there were fewer formal chiefs, but, paradoxically, more people willing and able to become 'acting' chiefs as and when required. But the irony is that while the German army was much better equipped to replace dead and injured leaders in the field, their lines of authority at senior level were so deeply bureaucratic that the Wehrmacht's top hierarchy remained implacably inflexible.

In fact, Allied officers were prime targets for German snipers throughout the European campaign because, as a German report recognized: the other ranks, especially, but not only the British, were so dependent on their officers – and their NCOs were so 'rarely in the big picture – that if the officer became a casualty, they were unable to act in accordance with the main plan ... The conclusion is: as far as possible *go for the enemy officers*' (quoted in Hastings 1993, p. 177, original emphasis). The selection of leaders for killing seems to have been standard policy from the very first minute on D-Day. For example, the official history of the 29th Division's landing on Omaha suggests that:

> within 7 to 10 minutes after the ramps had dropped, Company A had become inert, leaderless and almost incapable of action. The Company was almost entirely bereft of officers ... Every sergeant was either killed or wounded. It seemed to the others that enemy snipers had spotted their leaders and directed their fire so as to exterminate them.
>
> (Quoted in Holt and Holt 1999, p. 108)

Brigadier Hargest , the commander of New Zealand's 5th Infantry Brigade who was captured in Libya, was of the same opinion – British officers were easily recognizable, especially by their shiny map folders, and without officers British troops were often unable to move forward or take the initiative on their own.

> The English soldier ... demands leadership by officers. I notice that, as soon as men lose their officers, they lose heart. Today, the 16th, I met several Bren gun sections coming back and in reply to my enquiries 'why!' they said the fire was heavy and they got out of touch ... It did not occur to them to go on ... This is not suggesting that they lack courage, they don't.
>
> (Graham 1999, p. 159)

To immobilize the American, and especially the British, army therefore, one only needed to kill or injure the junior officers, but to immobilize the German army one only needed to keep the senior officers and political leaders alive. The consequence of the German policy led many officers to begin 'making themselves as inconspicuous as possible, removing epaulettes, and discarding binoculars, clipboards and other tell-tale signs of authority and carrying rifles instead of revolvers' (Whitehouse and Bennett 1995, p. 28).[1] References to ranks were also removed by the adoption of codenames. For example, the commander of any formation – of any size – was referred to as SUNRAY and the second in command was SUNRAY MINOR (Forty 1998, p. 107). Saluting, calling officers 'sir', and so on, all fell by the wayside in many US units at the front as officers learned that survival was preferable to status (Balkoski 1999, p. 219).

In the US army combat officers often removed collar insignia but retained a vertical white stripe down the *back* of their helmets to aid identification for their own troops without drawing the attention of enemy snipers – though some officers covered the stripe with mud when it became clear that German snipers were often left behind departing German troops to harass the Americans from the rear (Balkoski 1999, p. 219). Likewise, Allied tank troop commanders initially led from the lead tank, often with their heads out of the turret. But the ability of the Germans to destroy the lead tank – and with it the co-ordination of the troop – quickly led to all tanks operating with the hatch closed and with the commanders travelling in the second or third tank with every other tank taking it in turns to travel at the front, on 'point' duty (Cooper 1998, p. 73).

The result of this polarity in the German command system was that German troops in the field almost always proved more than a match for most of the Allied units facing them. As one SS officer recalled: 'I felt when we were in the field that in a fair fight with all things equal we would always win, not because of superior bravery but through better methods which I assure you were always – ALWAYS – designed to preserve lives, or at least to win a battle with fewer casualties' (in Blandford 1999, p. 130).

The 40 divisions (900 000 troops) of the Waffen SS (literally 'Weapons Protection Squad'), were the particular enemy that allied soldiers feared most and wanted to eliminate quickest, though almost all the units of the Wehrmacht were extremely efficient. In contrast, many German troops were sceptical of the fighting qualities of their American and British enemies. Hastings also admits that the situations were different: the Germans were fighting for the survival of their country whereas the Allies were undertaking an unpleasant but necessary task; the Germans did not have an apparently unending supply of troops or supplies which ensured that sooner or later victory would be achieved and, therefore, there was little need for heroics. And, most significantly, from 1943 the Germans were primarily fighting a defensive war where most of the advantages lay with the defenders – though clearly this had proved irrelevant for the French army in 1940 or the Soviet forces in 1941. Ambrose insists, however, that when the Germans went on the offensive against the Allies, at Bastogne and through the Ardennes in the winter of 1944, the American troops in particular proved more than a match for the German elite soldiers (Ambrose 1995, pp. 52–3). This is probably accurate, though by that time most of the elite and veteran German soldiers had already died, either in Russia or in Normandy.

Whatever the truth of the situational conditions may have been there is little doubt that, unlike the rest of the Wehrmacht, the SS were simultaneously feared and loathed by most Allied troops. The SS were under the Wehrmacht for operational matters but were not under the latter's disciplinary system, being responsible directly and only to Hitler. The Leibstandarte Adolf Hitler (LAH), led by Sepp Dietrich, were involved in the 'night of the long knives' in 1934 and, by 1943 had become the 1st SS Panzer Division Leibstandarte. Their approach to war could not have been more different from that of the typical Allied soldier. As one LAH captain put it:

> It was those defensive battles in Russia which I shall always remember for the sheer beauty of the fighting, rather than the victorious advances ... After a time we reached a point where we were not concerned for ourselves or even for Germany, but lived entirely for the next clash, the next engagement with the enemy. There was a tremendous sense of 'being', an exhilarating feeling that every nerve in the body was alive to fight.
>
> (Quoted in Reynolds 1997, p. 12)

The average British soldier seemed to see military service as very much a temporary interlude before resuming civilian life. According to the Army's own morale reports, 'the majority of soldiers ... were not interested in a "new Britain", nor the stirring speeches that political leaders made but were simply concerned that they should return to a decent house and secure employment in the post-war world' (quoted in Crang 1997, p. 71). That said, the vast majority of the civilian population remained supporters of the coalition government – at least after the nadir of 1942 when a further string of defeats had appeared inevitable (Badsey 1994a, p. 283). By D-Day British army morale was said to be at its peak yet a letter from Captain Alistair Bannerman, a British platoon commander, to his wife summed up the British soldiers' motivation:

> To soldiers ... Churchill's radio rhetoric sounds a bit embarrassing. They have no great faith in the new world, they have no belief in any great liberating mission. They know it's going to be a charnel house. All they want is to put an end to it all, and get back to civvy street, to their homes, their private lives, their wives and loved ones.

In this they were little different from their American colleagues: Private Branham in the US 116th Regiment thought along precisely the same lines: 'We wanted to go. This sounds crazy, but we had come this far, we'd been sitting in England so long, we wanted to get this thing over with and get the hell home' (both quotes in Ambrose 1995, p. 162 and p. 185).

Of course, not all the soldiers in German uniform in Normandy were German and the Allied planners had assumed that the presence of such non-German soldiers would ensure a *relatively* easy assault. Self-evidently, they argued, identity was critical to combat efficiency and the absence of ethnic Germans would inevitably lead to weaker defences. The real problem would come later, when the SS and Panzer units arrived over the first week. The planners were wrong on both counts. First, although some non-ethnic Germans proved unreliable, many did not, and a large proportion of the ethnic German troops were not 'old and sick men' but veterans of the Eastern Front who proved more than a match for their green opponents. Second, the position and speed of deployment of first-rate infantry, SS and Panzer units ensured that the 'lightening breakout' from the bridgehead turned into a prolonged season of storms.

Since few Germans assumed Normandy would be the invasion area it was generally regarded as a 'rest and recuperation' area for units on leave from the Eastern front (where a 3.3 million German army had suffered 3 million losses between June 1941 and December 1943) or a zone where second-rate troops could be left in control freeing up the better units for defending the Pas-de-Calais or in Russia (Ambrose 1995, p. 33). By the end of 1941 Hitler even began conscripting Russian prisoners-of-war into labour battalions to fight the Red Army. However, they proved so unreliable that they were transferred to the West at the rate of two Ost (East) battalions for every one (native) German battalion. The proportion of 'Russians' – that is non-ethnic Germans – in the West has always been disputed. Von Schweppenburg put the total proportion of 'Russians' at one-third of the entire 7th Army, though only 17 per cent of the infantry battalions were actually formed from such troops, though there were several *volksdeutsche* units made up of ethnic Germans from outside the 1939 Reich (Schlieben 2000, p. 98). Of the Ost battalions that Masters (1997, p. 180) came across behind Sword beach, almost all were infantry soldiers commanded by ethnic Germans, and all the artillery and mortar units were composed wholly of ethnic Germans. Nonetheless, so great was the ethnic diversity that eight different pay-books were necessary and the Nazi obsession with racial purity became miraculously stretched to include *Volksdeutschs* ('Racial Germans'), non-Aryan *Freiwillige* (volunteers) and *Hilfswillige* (auxiliaries). The 987th Regiment of the 276th Division, for instance, had 34 different Soviet languages. Lieutenant-General Karl-Wilhelm Von Schlieben, commander of the Cherbourg garrison, justified its surrender on 26 June 1944 partly by reference to the quality of his troops: 'You can't expect Russians and Poles to fight for Germany against Americans in France' (quoted in Badsey 1994b, p. 159). But this did not mean the defenders would buckle under pressure. Moreover, the Ost battalions had been informed that anyone retreating would be shot by his German NCO, while anyone captured by the Allies would be shot as a traitor (Gale 1995, p. 240). If anything, the ethnic diversity of the American troops was often greater than that of their enemies. For example, the US 94th Infantry Division had troops from 48 states, who could speak 24 different languages between them and of whom 843 were born outside the USA (including 134 German-born) (Prefer 1998).

> But there were some socially homogenous German units. The 12th SS Panzer Division *Hitlerjugend* (*HJ*), for instance, was raised virtually from scratch in June 1943 and was the first *Waffen* SS unit in action in Normandy. It was very effective and very large: over 20,500 people were within its ranks, so popular had been its recruitment campaign. As a consequence over 2,000 soldiers were transferred to the depleted *Leibstandarte* to make up its numbers.
>
> (Reynolds 1999, p. 14)

The 12th SS was supposed to be composed of youths born in the first half of 1926, which meant that the *HJ* division was not full of fanatical *children*, since two-thirds were 18 during the Normandy campaign and the rest slightly older (Reynolds 1999, p. 15). By comparison, the average age of American soldiers in Europe at this time was 26. Nevertheless, children were engaged in other units as soldiers towards the end of the war. But, as Reynolds (1997, p. 19) makes clear, 'What made the *HJ* division different was that it was *based* on eighteen year olds'. This age-based commonality undermined any regional allegiances that the members may have had in precisely the same way that the Zulu impis were organized (see Grint 2000). Equally important, the training provided was different from that given to the rest of the German forces – and markedly in contrast with the norm for the Allied forces. Physical fitness, character training and weapons training were the three critical activities and the three critical prohibitions were alcohol, tobacco and women. This spartan regime – and the similarities between the *HJ* regime and the Spartan regime are noticeable – was not, however, buttressed by the authoritarian obsession with discipline, 'square-bashing' and respect for authority so beloved of martinets in all conventional armies. Parade bashing may have proved of great utility either in practical terms for the practising of close-order battle tactics used up until the American Civil War, or even for the indoctrination of blind obedience that every conventional army allegedly operated on, but, as its commander Kurt Meyer recounted, the *Hitlerjugend* was no ordinary army unit:

> There was no dominant superior relationship, recognizing only orders and unconditional obedience. The relationship between officers, NCOs and other ranks was that between older experienced and younger comrades … The boys were educated to a sense of responsibility, a sense of community, a willingness to make sacrifices, decisiveness, self-control, camaraderie, and perception ... During their training square bashing was frowned upon … Everything focused on training for battle and this took place under the most realistic conditions possible.
>
> (Quoted in Reynolds 1997, p. 20)

Even conventional Wehrmacht units minimized distinctions between enlisted men and officers – unlike the American and British armies in particular. In the Wehrmacht everyone was *encouraged* to salute everyone else, irrespective of rank – in sharp contrast to the Anglo-American forces where soldiers were *required* to salute officers. Indeed, while the term 'soldiers' covered everyone in the Wehrmacht, there was no term for everyone in either the American or British armies, there was just 'soldiers' and 'officers'. Could it ever be imagined that British or American officers involved themselves in something that German officers regularly did – despite the obviously and radically subversive consequences for military discipline – singing 'Happy Birthday' to their 'common' soldiers? Since both the British and American officers were formally warned against fraternizing with their own soldiers we can rest assured that such a thing was unlikely (Kier 1997, p. 30).

Training for non-SS troops comprised 16 weeks of elementary training and three more weeks of combat training but the whole thrust of that training was to reinforce the importance of using initiative. As the *Truppenführung* (Troop Leadership) manual insisted: 'Orders should be binding only in so far as circumstances can be foreseen' (quoted in Balkoski 1999, p. 68). Training for SS troops was tough, and all officers were expected to serve for two years in the ranks before becoming officers, but the enormous casualty rate incurred by most SS (and indeed most Wehrmacht) units led to a rapid turnover in staff and a consequently very young officer corps. In fact, so rapid had been the construction of the *HJ* that when it went into battle just after D-Day in Normandy over half of its soldiers had been members of the division for less than one month and most had been in possession of their weapons for less than this time (Reynolds 1999, p. 14). In effect, the *HJ* was no battle hardened or even trained unit – but it was ideologically committed and it was extraordinarily effective.

The problem for Allied soldiers was not that they resented the idea of a military system (for most recognized its functional requirement) but many resented the way the disciplinary system seemed more intent on humiliating them than on organizing a combat unit. The Anglo-American commands followed conventional military doctrine in assuming that the only way to create and maintain a mass army was through mass coercion, to the point where individuals did not think for themselves but reacted instantly and positively to whatever command they were given. For different reason both the UK and the USA adopted very similar training methods. In the US the military establishment deemed the problem to be the endemic individualism of US culture, but for the British the problem was configured as one relating to the 'collective individualism' which became manifest in British trade unions, a utilitarian adherence to self-protective collective resistance. The solution to both problems was the same: the Anglo-American armies would administer 'intensive shock treatment' that would render the individual helpless before the institutional power of the military, and this would then allow the latter to rebuild the former in their own images. As Henry Giles quickly realized on joining the US Army: 'The first thing the army teaches you is if you've got any brains it's best to keep it to yourself' (quoted in Linderman 1997, p. 186). This shock treatment would force the individual to recognize that the group was critical not the individual, and that the group's role was not to protect its own interests but to advance the interests of the body politic and its substitute authority, the army.

The result was a British 'bullshit' and American 'Chickenshit' system that saw soldiers digging 6-foot graves for a carelessly dropped cigarette, cleaning toilets with toothbrushes and endlessly perfecting marching techniques on the parade square. The 1926 British infantry training manual, for example, had just two topics: (1) 153 pages on various aspects of drill, (2) 64 pages on 'ceremonial' (Kier 1997, p. 125). Luckily for the Germans there was little or nothing that resembled combat training, but then that would all be taken care of by the orders that would be given by the officers. They, in turn, having been thoroughly versed in regimental tradition, would be naturally capable of doing whatever was necessary to maintain that most important winning formula: morale. The 1924 British cavalry training manual was little better: of the 377 pages, 212 pages (56 per cent) were devoted to equitation, 96 pages (25 per cent) were devoted to drill and ceremonial, and only 36 pages (9 per cent) to weapons and field operations. The contrast with the Royal Air Force where professional and technical expertise was all and disciplinary perfection nothing, could hardly be more complete (Kier 1997, pp. 126–7).

This is not to say that the allies did not have ideologies, or that they were mechanically driven by coercive systems in the absence of ideological cohesion, but rather that the ideological cement that kept the allies together was less important and more internally variegated. The allied officers may have been more ideologically committed than their soldiers, but the same soldiers were committed to each other and to their mutual survival.

It should also be remembered that the traditions of the German industrial system were also in sharp contrast to those of the Americans in particular. German industry had concentrated on low-volume, high-value products – for which considerable technical expertise (*Fachkompetenz*) was a universal requirement – and which led to a system whereby industrial leaders had to work their way through the ranks to achieve their position. The same philosophy underpinned the German army: potential officers had first to undertake six months' basic training along with all other recruits and then train with their regiments for one year as private soldiers. Even during the war when officers were in short supply no German could achieve officer rank without first serving for two months in the line as an enlisted soldier. The result, as even the Military Intelligence Division of the US War Department admitted, was that in the 'practical arts of war', that is, in actual combat, German units fought, and were almost always led, much more effectively than their opponents (Locke 1996, pp. 67–70).

American industrial policy had been the reverse: concentrate on high-volume low-value products using mass production and machinery that obviated the need to rely upon skilled workers. Hence American industrial leaders could achieve their position without going through the ranks – either in industry or in the military. Where training concentrated on organizational 'systems' the result was an American superiority in production, logistics and supply train that almost always out-produced and out-supplied their German opponents, but were seldom their equal in ground combat.

The British establishment, despite their early industrial lead, had always been more interested in the accoutrements of high social class rather than developing a cadre of efficient and effective industrial management. So, paradoxically, while the background of the German army was not dissimilar to the British (they were both deeply rooted in the aristocratic land-owning classes), the British Army persisted in regenerating an aristocratic culture within the cavalry and infantry officer corps, while their Prussian counterparts adopted radically different practices and ideologies that, in some ways, inverted the conventions of the aristocracy. As a consequence the hierarchy and authoritarianism so prevalent in the British Army was displaced by a far more egalitarian and individualist orientation in the German army. As Kier argues, it was not a case of mechanically relating the class origins of officers to their professional culture because the class backgrounds of the British and the German armies were similar, but their professional cultures markedly dissimilar (Kier 1997, pp. 150–51). Indeed, the German navy was less aristocratic in origin but more aristocratic in culture than the German army, and even when the British Army became less dominated by the aristocracy, it still maintained its aristocratic culture.

This shift in German military policy from the traditional *Befehlstaktik* (an uncompromising order) to the new *Auftragstaktik* (mission description and goal sought) also embodied the essence of the division between the German and the western Allied armies for, ironically, it was the liberal democracies that most clearly relied on *Befehlstaktik* not *Auftragstaktik.* As Lucas (1998, p. 25) suggests, the *Befehlstaktik* principle was clearly present when the US forces landed at Anzio in February 1944, under orders to land and prepare for defence. They did precisely as ordered, even though the German defenders were caught unawares, leaving an opportunity for the exploitation of a local contingency that most German commanders would have found impossible to resist had they been the invading force. As Churchill concluded, the Allies had intended to throw a 'wildcat' into the belly of the enemy but instead they had discharged a 'beached whale.' (quoted in Murray 1995, p. 323). Such defensively minded 'inactivity' was not a consequence of anything other than a culturally conditioned response to what appeared normality. With disarming candour John Hogan, from the US Army's 9th Amphibian Force, captured this perfectly: 'If you haven't got stripes, you aren't doing anything … Every day I see the problem mounting and more and more maladjustment building up' (quoted in Linderman 1997, p. 53). Hargest, an acute analyst of the Allied armies, was also clear about the difference between the German and the Allied armies. It was certainly not lack of courage that distinguished them because the casualty rate of those that led the Allied troops – their field officers – was alarmingly high. But that was the point, if the officers did not lead from the front and expose themselves to enemy fire little could be achieved. 'On the German side, on the other hand … they were capable of carrying out their tasks without the continued presence of immediate leadership' (quoted in D'Este 1983, p. 281).

On the eve of the Second World War, and after Hitler – the arch-authoritarian whose *Führerprinzip* embodied the *Befehlstaktik* principle – had already spent three years in control, the official army manual repeated the approach that von Moltke had enunciated many years before: 'A favourable situation will never be exploited if commanders wait for orders. The highest commander and the youngest soldier must always be conscious of the fact that omission and inactivity are worse than resorting to the wrong expedient' (quoted in Kier 1997, p. 150).

That subordinate initiative became apparent at many different levels: subordinate officers replaced their dead or injured superordinates as quickly as they were themselves replaced by their own NCOs, who, in turn, were often substituted for by privates. And it also enabled local commanders to organized 'battle groups' (*Kampfgruppen*) to respond to local emergencies – as indeed occurred behind Omaha beach on D-Day. But the opposite tradition – overt and total control from the top, as developed by the early Prussian army under Frederick 1st – which the German military theorists had spent so long to overcome, was critical to the undoing of German defences.

By 1944 Hitler had seriously undermined both the spirit and the principles of the German command system. In the old army, teaching on the subject of command depended upon a cold and sober assessment of every situation, and relied upon the competence of every subordinate to carry out his task in whatever manner seemed best to him. This was replaced by 'intuition' from Berchtesgarden, and by a strict control of every smallest detail from the top. Contrary opinions were not entertained.

Paradoxically, then, while the German ideology in its Nazi guise rendered the world beyond ethnic Germans as one full of subordinate races and sub-human species in a vertical hierarchy, the inward focus rendered all Germans as virtually equal as horizontally ordered in a heterarchy – a fluid hierarchy. But the official Anglo-American ideologies of liberal democratic capitalism that outwardly espoused tolerance, legal and political equality, human rights and individual freedom, simultaneously had an inward military focus that polarized the world between the officers and the rest, a polarization at its worst that counter-posed the privileged gentlemen to the coarse cannon-fodder.

However the fanatical devotion to duty by some Germans, while it spawned frighteningly efficient military units, also created the weak link that undermined the German military on D-Day in particular and Normandy in general: the defenders were capable of pushing the Allies back into the sea, but the surrender of initiative by their senior commanders in deference to Hitler's personal control over decision-making effectively immobilized them. After the battle, General Queseda remarked: 'One's imagination boggled at what the German army might have done to us without Hitler working so effectively for our side' (quoted in Hastings 1993, p. 212), and there is certainly some truth in this. But what Queseda missed was the irony of the German military performance: it was primarily *because* of their personal devotion to, and identification with, Hitler that the elite and SS units fought with such tenacity; thus their strength was simultaneously their weakness. Only if Hitler had been less irrational would their power have been adequately exploited by the Germans but it was only because of their emotional – that is, non-rational – allegiance to him that their power existed.

As for the German officers, their role was to become both the political and the tactical leaders of their troops, unlike the Soviet system where political commissars were distinct from the military leader (though both countries adopted the other's system as the fortunes of war were reversed from 1943). Thus, although much effort went into the construction of political indoctrination amongst the Wehrmacht, the greatest reliance was placed upon the active role played by the local commanders. In other words, the motivational thrust came primarily from the military not the Nazis. The effects of this ideological campaign, of course, are much harder to establish, and perhaps we can get no closer to the 'average' Wehrmacht soldier than to suggest, following Schulte (1997, p. 282), that there was no 'clear dichotomy between fanatics fighting a religious war, or ideology-free and perhaps bored and apathetic soldiers; but a continuum'.

To most American combat soldiers the war was not a crusade against Nazi Germany it was an unpleasant 'job' that had to be done. The US Army pamphlet, 'Army Talks' devoted just one page to the ideological crusade against the Nazis and the CO of the 2nd battalion the 116th Infantry (29th Division) was clear that what most of his soldiers and fellow officers were fighting for 'was not ideals but rather an overwhelming desire to kill Germans, to avoid being killed oneself, and to achieve the terrain objectives we had trained so hard and long to reach' (quoted in Balkoski 1999, p. 63). Even those elite troops who were more ideologically involved than their ordinary infantry colleagues were frequently motivated to return the situation to the *status quo ante* rather than create some new 'utopia' or more likely 'dystopia'. As one commando with the Inter-Allied No. 10 Commando suggested: 'We may not have known what we were fighting *for* but we sure as hell knew what we were fighting *against*.'

That the task was a 'job' both rationalized the killing and retained some form of link with their pre-war normality – though again for quarter of the employable population the previous decade had seen unemployment as the nearest thing to a job. Hence, however unpalatable the 'job' of soldiering was, it was at least a job of some form and it provided both financial resources and the source of pride for many that had previously lost everything. In the words of Henry Giles, an American soldier since 1939, 'Nobody knows what the army meant to me – security and pride and something fine and good ... (For) the first time in my life I was *somebody* (quoted in Linderman 1997, p. 49).

The notion of a job also enabled those not in combat roles to feel that they too were helping the war effort. 'Drivers, supply clerks, cooks, mechanics,' wrote Captain Earl Nelson in 1944, 'are all winning the war simply by doing their job' (quoted in Linderman 1997, p. 52). That many took pride in their achievements can be related to the extent to which those *not* involved in combat have often sought to 'reinterpret' their past so that they can present themselves to the world as legitimate 'veterans'. But few Allied soldiers had been through what most German soldiers went through.

CONCLUSION

The German forces were more ideologically coherent than the Western Allies and more effective. But that effectiveness, though, was also rooted in a set of strategic imperatives and field tactics that were associated with elite units and not fundamentally derived from the Nazi philosophy. However, ideological cohesion remains a double-edged weapon: it does appear to propel groups forward faster than less cohesive groups, but the direction may be problematic, especially if the support is uncritical, closer to a leader cult than a culture of leadership. This is not to say that organizations cannot be effective without such a consensus: both the British and American troops were solidaristic, rather than consensually orientated, and they proved effective enough to defeat their more ideologically committed opponents. Moreover, that solidarity was not rooted in a stable environment because that was what the troops had been removed from. However, the lure of stability and the return to normality seems to have been important in persuading the Allied troops to continue. But this was an ideology that participants *imposed upon* events rather than one which was *derived from* events.

The elite units of all sides operated through cultural codes that were always rooted in opposition to one or more 'others', and not necessarily their overt enemies. For example, although overtly hostile to the Germans, the British Commandos were also in competition with, and hostile to, the British Paratroopers. In short, ideological solidarity seems to be closely associated with boundary construction: the stronger the group boundary the more equity appears to prevail inside and the less outside.

Another important lesson from the war might be that a healthy scepticism about the *unique* significance of contemporary 'instability' might be the first prerequisite for the debate, because we appear to have always lived in a permanent state of instability when considered in any significant historical light. The review of the war in Western Europe suggests that the connection between instability and cohesion remains tenuous because it is not mechanically determined by the material circumstances. That, of course, also means that the perception of instability – and its profound effects – might be far more important now than during the war.

Cohesion rooted in ideological consensus is undoubtedly a powerful elixir for organizational success and it can be imposed upon a situation with considerable benefits to that organization. But it also carries the seed of its own destruction where the culture of leadership is transformed into the cult of the leader. As elite units and elitist groups have shown, the construction of a boundary against 'the other' remains a fertile strategy for those seeking to unite and mobilize a population, but the very fervour of that mobilization may be its undoing.

NOTE

1. From August 1941, with Soviet collapse assumed to be just weeks away, Stalin ordered that any Soviet officer found to have removed his or her insignia could be shot by other members of the officer's unit (Magenheimer 1998, pp. 95–6).

REFERENCES

Abercrombie, N., S. Hill, and B. S. Turner (1980), *The Dominant Ideology*, London: Allen and Unwin.

Ambrose, S. E. (1995), *D-Day June 6, 1944*, London: Touchstone.

Anderson, B. (1983), *Imagined Communities*, London: Verso.

Badsey, S. (1994a), 'Great Britain', in D. G. Chandler and J. L. Collins (eds), *The D-Day Encyclopedia*, Oxford: Helicon, pp. 282–5.

Badsey, S. (1994b), *D-Day: From the Normandy Beaches to the Liberation of France*, London: Tiger Books.

Balkoski, J. (1999), *Beyond the Beachhead*, Mechanicsburg, PA: Stackpole Books.

Blandford, E. (1999), *Two Sides of the Beach*, Shrewsbury: Airlife Publishing.

Colley, L. (1994), *Britons: Forging the Nation 1707–1837*, London: Pimlico.

Cooper, B. Y. (1998), *Death Traps*, Novato, CA.: Presidio Books.

Crang, J. A. (1997), 'The British soldier on the home front', in P. Addison and A. Calder (eds), *Time to Kill*, London: Pimlico, pp. 60–76.

D'Este, C. (1983), *Decision in Normandy*, London: William Collins.

Dupuy, T. N. (1977), *A Genius for War*, London: Macmillan.

Fox, A. (1971), *The Sociology of Work in Industry*, London: Collier Macmillan.

Fox, A. (1974), *Beyond Contract*, London: Faber.

Forty, G. (1998), *British Army Handbook 1939–45*, Stroud: Sutton.

Fraser, D. (1999), *And We Shall Shock Them*, London: Cassell.

Gale, R. (1995), '6th Airborne Division', in W. G. Ramsey (ed.), *D-Day Then and Now*, London: After the Battle Publications, pp. 214–49.

Goldthorpe, J. H, D. Lockwood, F. Bechofer and J. Platt (1968), *The Affluent Worker*, London: Cambridge University Press.

Goldthorpe, J. H, D. Lockwood, F. Bechofer and J. Platt (1969), *The Affluent Worker in the Class Structure*, London: Cambridge University Press.

Graham, D. (1999), *Against Odds*, Basingstoke: Macmillan.

Grint, K. (1998), *The Sociology of Work*, second edition, Cambridge: Polity Press.

Grint, K. (2000), *The Arts of Leadership*, Oxford: Oxford University Press.

Grossman, D. (1996), *On Killing*, New York: Little, Brown and Co.

Hastings, M. (1993), *Overlord*, London: Macmillan.

Holt, T. and V. Holt (1999), *Major and Mrs Holt's Battlefield Guide to the Normandy Landing Beaches*, Barnsley: Leo Cooper.

Janis, I. L. (1982), *Groupthink*, second edition, Boston: Houghton Mifflin.
Kier, E. (1997), *Imagining War*, Princeton, NJ: Princeton University Press.
Lindermann, G. F. (1997), *The World within War*, London: Harvard University Press.
Locke, R. R. (1996), *The Collapse of the American Management Mystique*, Oxford: Oxford University Press.
Lucas, J. (1998), *The German Army Handbook*, Stroud: Sutton.
Magenheimer, H. (1998), *Hitler's War*, London: Arms and Armour.
Masters, P. (1997), *Striking Back*, Novato, CA: Presidio.
Milgram, S. (1974), *Obedience to Authority*, London: Tavistock.
Murray, W. (1995), 'The world at war 1941–45', in G. Parker (ed.), *Cambridge Illustrated History of Warfare*, Cambridge: Cambridge University Press, pp. 320–39.
Pears, I. (1997), 'The gentleman and the hero', in K. Grint (ed.), *Leadership*, Oxford: Oxford University Press, pp. 231–49.
Peters, T. J. and R. H. Waterman Jr (1982), *In Search of Excellence*, New York: Harper and Row.
Prefer, N. N. (1998), *Patton's Ghost Corps*, Novato, CA: Presidio.
Reynolds, M. (1997), *Steel Inferno*, Staplehurst: Spellmount.
Reynolds, M. (1999) *Men of Steel*, Staplehurst: Spellmount.
Rodrik, D. (1997), 'Has globalization gone too far?', *Californian Management Review*, **39** (3), 29–35.
Roseman, M. (2002), *The Villa, the Lake, the Meeting*, London: Allen Lane.
Rowe, D. (2000), *Friends and Enemies*, London: HarperCollins.
Schlieben, K. W. (2000), 'Background: the 709th Infantry Division', in D. C. Isby (ed.), *Fighting the Invasion*, London: Greenhill Books, pp. 98–108.
Schulte, T. J. (1997), 'The German soldier in Occupied Russia', in P. Addison and A. Calder (eds), *Time to Kill*, London: Pimlico, pp. 274–83.
Standage, T. (1999), *The Victorian Internet*, London: Orion.
Whitehouse, S. and G. B. Bennett (1995), *Fear is the Key*, London: Robert Hale.

PART II

Personal Coping Strategies of Organizational Actors

6. Management as a Dead End: How Managers Survive in a Culturally Complex Environment

Frans Kamsteeg and Willem Koot

MODERNIZATION, COHESION AND MANAGEMENT

'Radical changes in contemporary society have set conventional relationships reeling.' These are the opening words of the NWO (Netherlands Organization for Scientific Research) programme on Social Cohesion (NWO 2000, p. 7). The programme goes on to cite sweeping changes in norms and value patterns, in the commitment and involvement of people in general and, last but not least, in the relationship of people to organizations where personal commitment and involvement were, until recently, more or less taken for granted. The predisposition towards cohesion in (our) culture is on the wane. At least, so it would seem. Here in the West we are witnessing a loss of solidarity and trust, though these are actually indispensable in a hardening society. In short, mutual ties and cohesion – or more precisely, the lack thereof – has become a social problem.

The Dutch government believes that social cohesion can be steered in the right direction and is attempting to improve our understanding of the problems by sponsoring the Social Cohesion Programme. But can social cohesion actually (still) be managed? The instruments of management are, to say the least, shrouded in obscurity. Moreover, this belief that social processes can be managed is shared by many. The traditional civil society organizations were set up primarily because they enabled alliances to be forged within the respective socio-religious sectors (in Dutch often called 'pillars'). Whether this is still the case is very much open to question. The Dutch government sees itself as responsible for safeguarding social cohesion and is attempting to do so by passing legislation, but also by supporting civil society organizations. It has become the guardian of a harmony model known as the 'polder model', which has even come to enjoy international renown. Finally, the business community also sees commitment to the organization as a valuable asset and a virtually essential precondition for an effective economic performance by the employees.

At the moment, however, the role of the government and civil society organizations as the organizers of commitment is in a seriously weakened state, and more and more businesses are organizing meaning and purport for their employees. What Habermas (1981) describes as the 'colonization of the life world' (in which he alludes emphatically to government meddling) is becoming increasingly evident in professional organizations. As has been argued so convincingly by Casey (1995, pp. 194 ff), more and more people are deriving their identity from their job and their professional output. This implies that the organizations they work for are also fast becoming crucial sources of identity-construction (possibly the most extreme form of colonization), making it less likely that other forms of commitment, solidarity and cohesion will evolve outside the professional sphere. So, we could be witnessing a general decline in social ties, which is being paralleled by a shift in the focus of commitment towards the professional organizations. If this is true, then the researchers of the business community and also of a government that is concerned about social cohesion should redirect their attention to the social role that is currently being fulfilled by these organizations and the significance of commitment within them.

But what has actually given rise to this perception that we are losing cohesion and commitment? Academics and the (popular) press are inclined to blame trends in (late-) modern society, which is frequently described as highly individualized, hyper-rational, differentiated and secularized (Beck, Giddens and Lash 1994; Giddens 1991; Van der Loo and van Reyen 1997). In actual fact, it all hinges upon radical changes in the patterns of cultural thought, or 'basic cultural assumptions' (Schein 1985, following on from the work of Kluckhohn, Strodtbeck and Roberts 1961), that dominate the late-modern era. Individualization, differentiation, domestication and rationalization, the main principles of the Enlightenment, have instilled in our culture a rational instrumental worldview, which provides answers to life's most urgent questions: how to relate to Nature, how to judge human character, how to define truth and reality, how to interpret human relationships … In the long run, what it all boils down to is a strongly individualized trend towards management-orientated and progress-orientated thinking. But the modernization process is far from simple. The consequences of the Enlightenment principles are themselves fairly paradoxical. Modernization and Enlightenment appear to represent human freedom, but there are plenty of arguments for postulating that independence, pluralism, fragmentation, conditioning, alienation, uncertainty and dislocation are just as much the results of modernization. Beck's *Risk Society* (1992) demonstrates beyond doubt the downside of our enlightened idea that we are always in control. So, essentially, everything that we thought had been eliminated by our rational–instrumental approach has reverberated upon us like a boomerang that may even have gathered added momentum along the way.

One of the more dire aspects of this modernization paradox is that social cohesion and commitments have declined as a result of this belief. We shall see that this paradox also manifests itself in the business world, and that our first (and possibly only) response is to cling even more desperately to the modernization ideals as a means of combating the very problems they create. Despite the negative aspects of these Enlightenment ideals, our faith in the manageability of social processes is still strong. This is hardly strange, given the obstinate nature of culture. Our basic cultural assumptions about the world around us and human relationships seem to be driving us – according to our hypothesis – to look for solutions to management problems in a repertoire of cultural actions which has been largely conditioned by rational instrumental philosophy. But this means that we must face up to this paradox. So, how do we 'survive' the effects of the illusion of manageability that we continue to uphold, even though we often know better?

The seriousness of this problem in organizations and businesses, coupled with the theory that people are increasingly deriving meaning and identity from the work they do, is making us concentrate more strongly on the paradoxical consequences of modernization in organizations. Whilst belief in the manageability and controllability of organizational processes is criticized and belied on all sides, the denial and repression mechanisms which are at work here are exactly the same as those employed by beleaguered religious believers. We shall present two cases to show how these paradoxical consequences of modernization manifest themselves in professional organizations, the problems they pose for the individual, how managers relate to them and the solutions they come up with in order to 'survive'.

LIMITS TO MANAGEMENT AND CONTROL IN PROFESSIONAL ORGANIZATIONS: TWO CASES

In order to understand the paradoxical nature of modernization in organizations we must first recap on how the dominant management discourse has developed in relation to management practices. We shall then explore how individual managers – a group of top managers in the Netherlands and a group of project managers in a Dutch organization for development aid – deal with the consequences of their own 'control ideology' and discuss the survival strategies that they employ.

Broadly speaking, 'management' is a multi-faceted term. It involves 'regulation, organization and steering' on the one hand and 'an ability to cope' on the other. These two aspects are, in fact, what we are referring to when we speak of the problem of 'dual (double) control' that confronts every manager. The manager must steer others, but must also steer him or herself. Both aspects have been present in the management debate from the start, but the emphasis has lain on the former for some time. Management tasks that were considered valid by Taylor and Fayol, such as planning, organization, issuing orders, co-ordination and control are still customary, but they have acquired a more subtle meaning, even though the means of realizing them remain predominantly instrumental rational. Organization experts have introduced no real changes to the discourse on control and management for decades. Gradually, greater value may have been attached to flexibility (Mintzberg 1979; Pascale 1990), decentralization (Wellins, Byham and Wilson 1991), culture (Deal and Kennedy 1982; Peters and Waterman 1980; Schein 1985), self-management (Merry 1995), employee empowerment (Block 1990) and the 'learning organization' (Senge 1990), but these concepts too are essentially management tools from above (cf. Alvesson and Willmott 1996; Casey 1995; Hope and Hendry 1996; Knights and Willmott 1999; Koot, Sabelis and Ybema 1996). This 'humanization' of the management discourse has been in evidence since the emphasis shifted to motivation and reinforcement of the self-image under the influence of the Human Relations school of Mayo (1930–40). This led to the discovery of informal processes in organizations, which have, in the meantime, only become more sharply defined. As a result, we are now seeing the emergence of an applied management of meaning, which requires not only the management of the business processes as such, but also of their meaning for the employees, one of the principal aims being to improve control of commitment to the organization.

All of this has made a deep impact on the individual manager. After all, in the culture and identity approach, which has been taking root since the 1980s, it is still primarily the managers who determine the mission, identity and core values of the organization through a form of meaning management that is designed to create unity and convergence. The manager had to be a charismatic, inspiring and coaching figure and be intellectually challenging as well – in short he or she had to be transformational – in addition to performing his or her more concrete tasks, even though these were adjusted in the transactional sense (Bass 1985). Schein (1985) even turns the manager into the founder of the – corporate – culture, who has to provide the organization with integration, stability and continuity (in a word, cohesion) to which all the members would more or less automatically subscribe. The belief that a unified culture can actually be made has lost hardly any of its credibility since the 1980s and 1990s. This leads to the following summarized conclusions:

1. The instrumental approach directed at control of the organization and its members has not been seriously eroded in management thinking during the past 100 years.
2. The opposite is true, however, of the means of realizing this control, which, in their refined form, cause increasingly more extreme configurations in the colonization of the life-world of the employees.
3. Consequently, the tasks required of the manager are making greater challenges upon his or her aptitude for self-management. His or her competencies are expected to include management and self-management, delegation and control, confidence and self-confidence.

The last point is especially important to the development of our argument now that the second meaning of management has evolved so strongly. In the Taylorian context the manager's task was still wieldable, for he or she was directly concerned with the control and management of concrete work processes. Under the regime of a strong, corporate culture, however, he or she became the culture maker, the culture bearer and the person who – as a symbolic leader – placed his or her trust in others to help them realize self-management within that culture. The expectations of and confidence in the manager as the general facilitator and communicator have, however, gradually become limitless. So, is the individual manager (still) capable of taking all of this on board? Can he or she handle this formidable and disconcerting task? How does he or she respond to it and what survival strategies does he or she employ?

And does the manager make the distinction between his or her role as a professional and his or her role as a private individual, or have both roles become hybrid?

Based on the data and material from the following two cases we can conclude that at the end of the day, the facilitating, unifying, encouraging and creative manager still appears to have a self-aggrandising, resolute, scheming and controlling alter ego.

Case 1: How Do Top Managers Survive?

The world of top managers has changed dramatically over the last 25 years. The former director, who was then fully in charge of his or her corporation, and was treated with great respect by his or her personnel, has now become 'just' a chairman/woman of the board. He or she is experiencing enormous pressure from the organization's stakeholders, amongst whom shareholders are rapidly gaining influence, thus reducing his former monopoly of power. The complexity of running a large corporation has also increased noticeably in today's globalizing world. Consequently, the manager is regularly confronted with localization and ethnic integration processes when he or she visits offices in other countries but is also confronted with resistance against his or her own head office. He or she has to create links with partners who, until recently, were either unknown to him or her or were his or her competitors. Moreover he or she must ensure that his or her complete organization exudes a certain level of uniformity, despite the inevitable local differences. Another complicating factor is that external conditions have become much less stable due to the process of virtualization and acceleration caused by the information technology boom. Growing (ethnic) diversity within organizations also contributes to the (feeling of) instability and insecurity. Finally, present-day top managers have to present and defend themselves and their companies publicly (in moral terms). They are therefore forced to account for their (companies') activities in societies that demand that they display civic responsibility and 'government with a conscience' (for example in the Shell Brent Spar affair). And, last but not least, the manager must show vigour and be able to plan well, and at the same time he or she should be 'warm' and supportive, stimulating and facilitative. So we can conclude that being a top manager in modern times means being able to combine many ambiguous and paradoxical skills and as such it has become an almost impossible task to fulfil.

Our first case addresses the way top managers in the Netherlands cope with the complexity of their task. The data in this paragraph is taken from a study by Koot and Sabelis (2002). They have focused mainly on highly successful managers who have been dubbed 'winners' by others and, not least, by themselves (in contrast to the people who appear in our second case). Despite this, these toppers experience a high degree of uncontrollability in the internal and external business processes that they are required to manage. But top management is a hard world, as they confided in Koot and Sabelis, where people can be severely punished whenever they express doubts about their own management ability. This makes it difficult for them to enter a discussion on the belief in the infallibility of classical management and management concepts.

This attitude emerged clearly at a weekend conference, organized by Koot and others, in the Ardennes, of top managers who did not belong to the sample who took part in the research. During the first session on Friday evening, the participants were given two lists of terms relating to corporate management. The lists were based on the material collected from the sample. The first list contained words such as scoring, realizing targets, control, planning, and creating corporate identity; the second contained words such as self-management, empowerment, the learning organization and diversity. It was suggested that these terms played a key role in the life of a top manager, even though they may seem contradictory at first glance. The managers were told that the words in the first list were used mainly front stage and that those in the second list were used mainly backstage (Goffman 1959). It was suggested they used all sorts of coping strategies to deal with the tensions triggered by this situation. One manager spoke for many of his colleagues when he fervently denied the existence of this paradox. During the course of the weekend – which was devoted to uncertainty and ambiguity for the present-day manager, a theme for which they had signed-up to with great enthusiasm – it gradually became apparent that the paradox sketched actually did reflect the daily dilemmas of many of the participants. While top managers enthusiastically proclaim that they are introducing a new, open corporate culture, they have often already prepared draconian restructuring programmes. They behave in this way because they are afraid of the consequences of openness, but they do nonetheless suffer from doubts and feelings of powerlessness and insecurity in the process. The conference ended with the conclusion that, completely unjustified, current management models pay no attention to the kind of dilemmas and ambiguities which arise in everyday management.

The cultural norm patterns which are revealed by these two discussions contain contradictions that could be described as the undesirable side-effects of modernization for the individual (Beck 1992). Managers experience their companies as unfathomable, and sometimes respond with harsh words against a background of mounting employee stress and the continuing decline of social cohesion. This is hardly surprising if managers advocate empowerment and flexibility (the slogan *be empowered* is, in itself, a paradox) but demand total loyalty and commitment at the same time. This is a situation that confronts people in senior positions in particular. Personal empowerment is facilitated; the individual has to show that he or she is always 'in control' (cf. Casey 1995). Whereas the dilemma of management and self-management used to affect only the leader/manager, it is now afflicting more and more individual employees. One of the results of this trend is that traditional ties are disintegrating and being replaced by strategic ties which are merely exacerbating feelings of uncertainty, confusion, ambiguity and lack of knowledge.

The paradoxical relationship between commitment and control which is articulated by the (top) managers in the above example is excellently explained in *Engineering Culture* (Kunda 1992), which traces the way in which the managers and employees of the computer company Digital are affected by normative control in a corporate culture. Kunda clearly illustrates how much of an instrument of control commitment has become in this type of management and describes the paradoxical implications that normative control can create. Here too, the management catchwords are culture, ownership, openness and so on, but behind the scenes the managers, and especially the personnel, point out that this debate is completely at odds with the performance norms upon which everyone in the organization, from top to bottom, will ultimately be judged. Culture and the unifying effects that it is meant to generate ought to be unmasked as a management tool, an ideology of control that, in fact, merely demonstrates how little has changed since Taylor's day. Empowerment and self-steering is encouraged, trust is extolled, but the underlying rationale is still instrumental and harnessed to control.

Managers say that managing their company in this way takes a heavy toll and that they are constantly plagued by feelings of uncontrollability. They are rapidly losing the self-confidence that pervades their rhetoric. How do they balance uncertainty and the pressure to perform? How do they cope with this dilemma? And how do they distinguish between their role as a professional and as a private individual? Essentially, in the same way as they manage their organization. The double control problem (Watson and Harris 1999) makes them manage others by managing their 'selves'. Kunda calls this an attempt to create an 'organizational self'. But whilst the manager uses the corporate goals and values to engender identification and commitment in his 'flexible' employees, he must also achieve (the impression of) self-management. This could be a tough task. A manager who has run out of solutions to problems of cohesion, co-ordination and commitment needs to suggest them. Indeed, we could say that (top) managers attempting self-management are confronted with the same sort of paradox as in the management of their companies. Front stage – in their working environment – they also suggest that they are in control; they ignore or deny the problem by employing cynicism/humour, macho behaviour – in fact, by tacitly upholding the rationale of self-management and self-control. This is almost invariably the first impulse.

These 'coping strategies' could be classified as emotion-regulators (Lazarus and Folkman 1984), designed to create a liveable situation. The (top) managers told the researchers about other emotion-regulating mechanisms, such as a dedication to art and/or (high) culture. Professional advice is sometimes called in to help select problem-solving strategies, most of which rely on the support of the partner/spouse/friends. Attempts are hardly ever made to change, redefine or restructure the situation. Someone mentioned the use of organizational models that deal specifically with the quality of human relationships. Only a few said that they spent time on (self-)reflection and self-disclosure (methods which are widely promoted, for example by Giddens 1991, for coping with ambiguities and paradoxes). Perhaps the scarcity of (self)-reflection can be explained by the fact that, as the study shows, most of these (top) managers are 'men of action'. They all said, with a few exceptions, that they simply do not get around to reflecting on the paradoxical consequences of modernization, even though the literature (for example, Merry 1995; Smets, Wels and van Loon 1999; Stacey 1996) suggests plenty of possibilities for allowing a 'mutual self' to emerge by using processes such as individual dis-identification, paradoxical thinking, showing trust or using other types of logic such as parallel circuits and holistic thinking. Incidentally, most of the managers confessed that they were unaware of these possibilities. To them, the complexity of their task lies mainly in the (daily) problem of management and control, rather than in commitment and cohesion. These conclusions drawn by Koot and Sabelis are very similar to those made by Watson and Harris (1999) for (top) managers in England. In England, we also see that the (verbal) emphasis 'backstage' lies in control and 'front stage' the emphasis lies in self-control, although, in this last case, the context of control is leading.

The top managers' non-reflecting behaviour is only logical if we view it in the light of Alvesson and Wilmott's hypothesis (1996) that reflection leads only to greater uncertainty. Why should managers who, in any case, do not have an abundance of self-confidence, as has become clear in this research, increase their uncertainty by reflection? They are more likely to opt for a form of impression management by aligning their identity with 'performance' (Bauman 1995) (compiling interesting CVs, embellishing their sporting achievements and so on).

The top managers who took part in Koot and Sabelis's research tend to make little distinction between their role as a professional and as a private individual. Most of them have adopted the 'impression management' method, which some of them have developed into a fine art. They also try to give the impression of having 'everything under control' in their daily private lives and give the impression of being wholly rational, efficient and goal and instrumentally orientated. An extraordinary example of this is the female top manager, Hannah (fictitious name), who has an office at home which is fully equipped with the most modern communications equipment and where she organizes teleconferences with her fellow board members all around the world. Although she is at home and no one can see her, she prepares herself meticulously for these conference calls. She ensures her make-up is perfect, wears a designer suit and sits in a disciplined position to give herself the feeling of self-control. Even in her doings with her family, being methodical and in control are uppermost in her thoughts. The worst thing that could happen is that she, the top manager, would become ill: 'that is dangerous, because then those around you will get the idea that you are physically vulnerable and that you have not got yourself and your environment under total control'. (For a further elaboration and an explanation of this amalgamation of work and private life see: Koot and Sabelis 2002.)

Even though the top managers know that it is impossible to have everything under control, in the reflection on their own private life they mostly allow themselves to be governed by the values control and rationality. They either ignore or dismiss the contradiction with the remark that 'in a while it will also be possible to control these processes too' or they divide it into a rational and irrational part, the latter quickly pushed into the domain of the 'soft sciences', 'women' or 'the church'.

There is plenty of evidence to indicate that though Dutch (top) managers do experience a certain degree of complexity in day-to-day management; they are seldom able to place this within a wider perspective. They cope with the pressures of management and self-management by appearing strong, by 'getting on with the job' (often intuitively) and by spending hardly any time on reflection. We could say that they are so preoccupied with management and control that their search for solutions is restricted mainly to this sphere; they more or less ignore the problems of commitment and cohesion, which we have identified as one of the central paradoxes of modernization. Or, more precisely, they use them purely as instruments ('normative control of organizational selves'). The conference mentioned earlier, in the Ardennes, clearly revealed where the greatest concerns of (top) managers lie. So, are the problems surrounding cohesion, commitment and identity more a matter of concern for the employees on the work floor, than for their (top) managers? Perhaps the second case will provide us with a few answers.

Case 2: Christian Organizations for Development Co-operation: Struggling with Identity, Planning and Control

This case is totally different from the previous one. Development aid organizations in the private sector cannot do a thing right. The people who work there are seen as 'losers' in contrast to the 'winners' in the study by Koot and Sabelis. Personal commitment to the mission of the organization one works for is seen as an obsolete ideal. And so it is to some extent. This sector stands, moreover, in the direct line of fire. Politicians who want to see results threaten it with financial cutbacks. There are serious doubts as to the value of diverse identities in the sector (Catholic, Protestant, Humanist, non-aligned). At the same time, more and more partner organizations overseas are talking back; they are no longer content to wait quietly for money, goods, experts or an ideological wish-list from the Western donor.

In the past, private sector Christian organizations that engaged in co-operative development aid operated in a domain that was far more neatly organized than it is today. Catholic organizations supported needy members of the Mother Church elsewhere in the world and the aid was accepted without too many ifs or buts. Moreover, as a member of a Protestant organization, you knew that you also belonged to the greater Protestant family, which formed a distinct part of Dutch society. Nowadays, however, these organizations are themselves no longer sure what their Catholic or Protestant identity actually means and they certainly have no wish to employ exclusively Catholic or Protestant staff, or to support only aid organizations from the same religious persuasion. Private sector development aid organizations in the Netherlands are now faced with the question of whether they can continue to maintain their time-honoured ideal of controlled development. Given the vulnerability of these organizations, it would be logical to expect that their project managers would be more likely than the (top) managers to have found alternatives for the belief in control. But this is only marginally true. Before we can demonstrate this we must first provide a rough sketch of the field.

Private sector development aid organizations in the Netherlands spend around €300 million of government money. The use that is made of this money is verified and approved every year through the Ministry budget. The sector is organized along religious lines. Hence, since the 1960s, it has comprized the ICCO (Protestant, interdenominational), the NOVIB (non-aligned/Socialist and the most well known of the organizations), the CORDAID (Catholic, founded in 2000 from various mergers between Catholic groups), and the HIVOS (humanist). Besides the money allocated by the government, these organizations also receive considerable funds from private sources, which is also spent on development projects. They all claim that they reach groups of people in developing countries which local governments and international donors, such as the World Bank, cannot or do not want to reach. They usually work in partnership with intermediaries, that is non-governmental organizations (NGOs) which run projects for poor population groups. Often (but not always) these NGOs share the philosophical/religious beliefs of the donor organization. In this section, we will specifically address the private Christian donor organizations.

Like the government, the private sector organizations have a reputation for generosity, but also for 'always knowing better' about development matters. The tendency to manage and control development processes from the Netherlands has always been very strong – the question of how people in other parts of the world *ought to* develop topped the list of policy plans of every newly appointed Minister. Though the objectives changed in the debate on development (for example, political awareness, combating poverty, women's rights, environmental education), development workers still clung to the idea that there is a useful (and urgent) message to bring on how development should be managed elsewhere (see Kamsteeg 1998). Over the years, the link between the results of development management and the invested funds has continued to strengthen. The Netherlands has always been a land of merchants and churchmen. This is still evident, especially in private sector organizations for development aid, where the clergy are sometimes actually running the show. Since the late 1980s, however, the politicians have been demanding more results in the form of concrete verifiable effects and hence also accountability for the taxpayers' money. Development workers are experiencing tension between 'all this bookkeeping stuff' and the pursuit of 'good works'. The present Minister, Herfkens, is also placing a heavy emphasis on verifiable results, control and financial management. Accordingly, she recently scrapped numerous countries from the donation list due to the absence of good government.

Let us now take a closer look at the practices of these 'ideological organizations' in order to get a clearer understanding of the complexity facing both them and their staff. We shall attempt to ascertain the kind of work the central players do and the dilemmas they experience. Our perspective will again be the dominant instrumental–rational approach to management, which we distinguished as the fundamental basic cultural assumption in our Western society. This is possibly best exemplified by the famous notorious project cycle in private sector development aid. The core business of private development organizations consists roughly of processing, submitting, approving, funding, monitoring and evaluating long-running projects. In this cyclical process criteria are constantly tested and the anticipated income (in terms of development) is placed against expenditure. The income/expenditure analysis, however, has always been a bit suppositious in development co-operation. The sector in general has concentrated preferably on objectives. In the private segment these were dominated by political and socio-cultural considerations that aid should contribute to the awareness processes of underprivileged groups so that they could improve their political participation and economic progress. The specific social groups that were eligible for subsidies were selected on the basis of their political/religious affiliation with the donor.

The discussion on the verifiability of these largely intangible objectives resulted in an increase in the number of reports on effects in the early 1990s. These reports revealed a wide discrepancy between the set objectives and criteria, and confirmed that it is extremely difficult to measure effects. This serious criticism exposed a paradoxical management problem. On the one hand, huge efforts were being made to steer the organization in the direction of the desired changes (the whole project cycle was, after all, geared to this); on the other hand, the context in which the interventions were taking place turned out to be so complex that the results defied management and therefore could not be verified. This paradox has actually always existed but it has manifested itself more sharply since the mid-1980s when good business practices – the demand for efficiency and effectiveness – ruled the roost in a strongly changing political climate. It could be said that the economic domain was making new and additional demands on the business operations of the sector. Project managers are feeling this pressure more and more in everyday practice.

At the same time, the identity of Christian organizations has undergone radical change. In the 1960s and 1970s development aid brought prestige. Aid workers were respected. The fact that the field consisted of religiously based groups which each served their own network in the Third World was not an issue. After all, a large proportion of Dutch society was arranged in the same way. Church affiliations were fairly normal among the employees and were even, to some extent, a precondition of employment. The organizations were also incorporated in Catholic and Protestant networks in the Netherlands (and internationally). This applied likewise *mutatis mutandis* to their members. All these fixtures have, however, rapidly disappeared. Secularization – which is closely related to the individualization and domestication aspects of modernization – has brought about a sharp drop in religious affiliation among the personnel. New employees are no longer required to show traditional commitment. Many of them see the Christian identity of the organization as a rudiment from the past that has ceased to be a binding factor. Bearing in mind also the current ascendancy of economist thinking, what all of this means is that the viability of the sector is at serious risk. How are the project managers responding to and coping with this situation?

Developments in recent years have shown that the project managers in their functional role have three main options which, though intrinsically different, still testify to a constant factor when rational instrumental aspects come into play.

Some of the project managers, who still have strong ties with the mission field, are arguing strongly for the continuation of the well-tried concept of managing development by, for example, sending out doctors, supporting relief organizations and so on. Most of the organizations that they work with are in the church domain. Others are pushing for projects where the possibility of tighter controls on efficiency and the quantitative measurability of effectiveness (results) are supposed.

Both these approaches fit in well with the trend envisaged by the directors towards 'accountability' thinking. Project managers must deliver results, verifiable results in the battle against poverty. In order to achieve this, the checks and balances in the project cycle need to be increased and improved.

There is, however, a third response that is being explored in some places. It involves the design and deployment of an alternative development model that provides for accountability and, at the same time, defines the Christian identity of the organization. At the Catholic organization CORDAID this means, for instance, that in the Latin America Department the principles of subsidiarity and accompanying actor-specific practices borrowed from the Catholic social doctrines are being developed in such a way that the donor's role is to facilitate qualitatively the sustainable development initiatives of the local players (basis organizations).

Also this third form of intervention is still largely directive: the efforts are designed to strengthen the position of the local players in the social dynamics and to generate social cohesion that will continue to exist, even when the donors have left. The project managers who work in this way are also pretending to breathe new life into the discussion on the identity of the entire (Christian) organization, whose legitimacy in this secularized age, devoid of tradition, is no longer self-evident. The way in which the Christian identity of the private development organizations is currently expressed is significantly 'narrower'. It revolves upon elements such as 'inspiration', 'tradition' and references to the Christian identity of the *partner* organizations.

When we examine all three responses from a broader perspective they appear to share a number of strongly rational instrumental characteristics:

1. The idea of intervention in project policy is still largely intact.
2. Accountability appears in the refinement of the checks in the project cycle.
3. The targeted end result, be it measured quantitatively or qualitatively, is invariably formulated along with a tightly regulated step-by-step plan that must deliver this result.

This donor driven, instrumental reasoning does not appear in the policy documents of the Christian development organizations. However, these documents do emphasize the capacity (and initiative) of Third World people and their organizations to solve their own development problems and to choose the appropriate means to tackle them (ownership). However, this must be considered principally as a front-stage discourse. Backstage, daily management practices remain predominantly directive and result driven.

In their role as employees of a private, Christian development organization the project managers show surprisingly little initiative. The increased external political pressure on the organization and the loss of a clear and widely borne identity may indeed lead to changes in policy, but this does not shift the fundamental belief from its central, guiding, position. Even a procedural approach, with the emphasis on self-management in the development projects, is kept in place by the high degree of control measures. The managers involved are apparently unable or hardly able to change their way of thinking. At the same time, as individuals, they are increasingly caught up in the disharmony between the theory and actual daily practice. How do they cope with this?

In practice, the working methods of the organizations, irrespective of their form, create a gigantic workload for the project processing, just to arrive at the best possible results. This pressure is merely exacerbated by the permanent feeling that the reality that one is trying to improve is not all that easy to manage. When this complexity reappears in the policy, as in the case of the actor-specific policy, it is experienced by many as unnecessarily complicated and time-consuming. It is as if the project managers are trapped inside the mechanisms they want to use for steering development processes in the field (the Third World). The sense of pointlessness is aggravated by the feeling that if anything is achieved then it will still be just a drop in the ocean and sometimes merely the outcome of unplanned interventions. 'What are we doing here?' is an often heard complaint. Many experience the absence of a clear relationship between objectives, tools and verifiable results as a fundamental problem of their job.

How can these project managers come to terms with all of this? Workload, stress and uncertainty (also about the future of the sector) are immense. The response to solving the problem is to 'make an extra effort'. After all, many project managers derive their greatest satisfaction from shepherding their own project proposals through the project committee – and from contacts with and in the field, especially on official trips.

Although project managers are constantly aware of what they are expected to do, namely manage development in situations where management is scarcely possible, very few of them dare to let go of this reflexive response and try out a more open ('process') approach to management which is less geared to preset results. Reflexivity is considered highly dangerous in so far as it could paralyse the work process, since it very easily leads to relativism. Relativism is, however, amply practised at the backstage level. Development workers are quite well known for their capability of combining utmost seriousness with blatant cynicism which is almost invariably related to the discrepancy between pretensions, plans and (organizational) reality. 'If the organization has no clear vision then why do we want one?' 'We're sitting here and talking about quality while in other departments or in other Christian organizations these kind of projects might have been approved much easier?'

This mixture of stress, compromise and cynicism in the daily routine only very rarely permits reflection on common working practices, the quality of project proposals and the feasibility or desirability of objectives. For many of them, there is no time, no scope and no need to reflect. Project managers who are pushing for an actor-specific policy do take some time to reflect, but they repeatedly run up against resistance. ('No, not that discussion on policy lines again, we simply have to get on with the job of delivering projects.') Project managers in private sector development aid are similar to the (top) managers in the previous case in so far as they too are primarily 'men of action'. They, too, fear reflection because reflection can make them realize that they have to let go of the hankering to control and that the core business is unachievable.

Back at the office, however, this leaves them searching for workable solutions to improve the quality of their management. The shortage time is a constantly itching problem. The option of delegating project management to a selected number of Third World partner organizations (NGOs) is seriously considered, but the project managers also start to push for better internal working conditions. Whereas, in the past, an appeal to the zeal of the aid worker was enough to spur him or her into going the extra mile, overtime is now regarded as far less normal. An interesting new development is taking place here in the relationship between work and the project manager's private life. Until recently the divide between these two domains was almost non-existent: the personal identity coincided almost completely with that of the development worker. The personal ideology to strive for a better world coincided with the content of the job within the organizations which also strove for this ideal. It was therefore easy to efface oneself for the higher (work) goal; after all, it was not seen as self-effacement. At a certain point in time, project managers tried to escape the ever more visible devices of the work by refusing to do overtime and to increasingly exercise their rights as employees above their assumed commitment to the organization.

To a certain extent, this partial un-identification with the organization also offers the opportunity to look outside the own organization. Until recently, the development sector resembled a lonely island. Once you were on it, it was rather difficult to leave. You had become addicted to the beauty and nobility of your work, while if you tried to escape from the island, you were received with suspicion and even hostility as your capacity to perform well in another working context was doubted. Development people were considered incapable of holding 'normal' jobs. Recently, however, the boundaries between this sector and other working environments have become more fluid. People have started to leave the sector and they prove to fit in well in other positions, exactly because of their usually well-hidden qualities, which they themselves are the first to deny, namely their management and control qualities.

Those who remain, seek relief in exploring new directions for identity formation. Although social cohesion in these Christian organizations is getting weaker and discussions on identity inspire little enthusiasm, a minority seeks to survive by relighting the fire. Commitment to the original religious character of the organizations may have more or less disappeared, partly because the personnel, including the project managers, have very little affinity with it. But there is still a Christian family to which these organizations belong, but this is only important to a few individuals. As such, it has only limited bearing on social cohesion of the organization's employees. One could possibly say that organizational identity has turned from a collective survival mechanism into a personal lifeboat for some of the Christian development workers.

Surveyed from this perspective, the thriving midfield organizations may be proof of the bankruptcy of the religious 'pillar model' in the Netherlands. What remains are organizations which are Christian in name only and which hide a multitude of diverse social and, above all, personal identities within their ranks, based on regional or thematic specialization. This certainly does not mean that potentially binding factors such as organizational identity do not have a future. However, it does mean that these are not the obvious choice today, and therefore much stronger factors are needed than those which tradition has provided.

CONCLUDING REMARKS

The time has come to draw a few tentative conclusions, formulate hypotheses and suggest openings for further research. We have discussed the paradoxical aspects of management and commitment from two ostensibly entirely different perspectives: the world of (top) managers in commercial organizations and the world of (project) managers in Christian organizations for development aid. Both groups see management and control as inherent parts of the job, but also as virtually impossible to achieve. The reality that confronts them cannot be 'tamed' or domesticated. The modern ideal that this is possible is repeatedly negated in practice. *Top managers* have no clear-cut conception of, or grip on, the fragmentation processes taking place in their own company or the increasingly complex environment. Supposedly liberating normative controls in the form of 'culture' and 'self-management' are to no avail. Time and again, they fall back upon trusted, more utilitarian forms of control (Etzioni [1961] 1975). This suggests that they also regard cohesion as predominantly functional. For them, cohesion is compliance with rules and regulations so that economic objectives can be pursued. So, at the end of the day, commitment is essentially a management problem. *Project managers* realize – though they do not openly admit it – that their ambitious development goals are often unachieved because they necessarily imply cultural changes which are largely unpredictable. Normative control has always been standard practice in the development aid sector. It inspired enormous enthusiasm that was fortified further in Christian organizations by the religious impetus behind the work. But the tension between enthusiasm and powerlessness is creating an irreconcilable dilemma. The reflex response to this dilemma is still the systematic approach (the project cycle with all its checks and balances). Modern innovations, such as the actor-specific approach, also have their own forms of control and management. What is more, the religious tradition which led to the emergence of these organizations is no longer a binding factor for either (the majority of) the people or the policy.

Both groups of individuals in our cases are running up against the paradoxical consequences of modernization. The liberating ideals of the Enlightenment have resulted in a fixed belief in the possibilities of management and control. The idea that an individual (top manager) can manage a large organization stems from this way of thinking, but – almost inevitably – it also becomes bogged down in it. The belief that development aid managers can develop the world has its roots in the self-same ideals, but it is relentlessly undermined in practice. The entrenched trend towards control in our Western culture (Arendt 1958) is driving people to break down barriers but, in their frustration, all that they do is collide with them. Differentiation, another characteristic of modernization, leads to specialization, but also to loss of cohesion. Our (top) managers try to restore cohesion by engaging in a debate on (cultures of) trust as the lubricant/cement of an organization, and of self-management and empowerment. In the development aid organizations, the project managers have lost the religious tradition as an ideology and a normative binding factor. Some of them try to reformulate this tradition, but the tide of secularization is against them. Their pillars of faith have turned out to have clay foundations, which are weakening social cohesion and causing the management practices to fragment.

The individual coping strategies are not particularly surprising in the cases we have cited. *Top managers* survive in various ways; they get support from friends, partners and sometimes even resort to professional advice, but they seldom opt for reflection. The options which management theory offers for redefining the situation by, say, dis-identification (Merry 1995), paradoxical thinking, trust and associative logic are wasted on them, apart from a few exceptions. It seems that some of the managers still cope by 'building' their identity on performance (Bauman 1995), which provides more in the way of image than of substance. Additionally, they can, in their position, still fairly easily fall back on the exercise of power – which is then also immediately subservient to control. Being, first and foremost, 'men of action' rather than philosophers, they are less inclined to reflect. But this may be because they are afraid that reflection will only make them (more) uncertain. It would not be unreasonable to suggest, then, that surviving in a culturally complex organizational environment may not really represent such a major problem for them. Indeed, it may be greater for the *development aid managers* in our second case. Their seemingly everlasting cynicism has gradually worn off; the hectic planning surrounding official trips is eroding their faith in the usefulness of maintaining contact with target groups in the Third World. Public respect for the sector is also declining. In our current flourishing economy the work that was inspired by the enthusiasm of the 1960s seems somewhat out-dated. It is neither fast nor cool. In other words, it is for losers. Development aid managers are not philosophers either, at least not in the reflective sense. They concentrate mainly on planning and implementing projects, which consume – or which they allow to consume – all of their time. As was revealed at a study day organized last year by the Foreign Office, it is symptomatic of a limited capacity for reflection that academic studies on development aid are not read and – worse still – are even regarded as more or less irrelevant for practical purposes. All in all, it is hardly surprising that many project managers feel threatened by stress or burnout (if they have not already succumbed). When this happens, coping is no longer in the picture.

Our cases pose two fundamental questions. How far have we become prisoners of our culture of control that began with the Enlightenment and finds expression in the principles of modernization? Are there any signs that we can escape this, that changes are taking place which are causing cracks in this profoundly obstinate culture? Are not the changes in thought patterns and working methods triggered by the computer (for example, complexity theory and new, non-linear thought trends in informatics) just as revolutionary as those triggered by the telescope, which Arendt sees as the model for the modern culture of enlightenment and control – a culture whose (bitter) fruits we are now plucking?[1] And, secondly, actually how necessary and desirable is commitment?[2] Following on from Kundera, Alvesson and Willmott (1996) say that it is unendurable to live without commitments, whilst at the same time commitment itself becomes a leaden weight. Or will we gradually end up subscribing to the post-modern form of commitment (for example, Bos 2000) with temporary but intense ties and loyalties as described by Maffesoli ([1988] 1996, pp. 139 ff.)? This is, at any rate, a trend that is recognized by some of the more reflecting (top) managers, and is also seen as a welcome and viable alternative. Empirical research could provide an answer to both these questions but it is still in its infancy. A possibly important factor that should be taken into account here is the impact of the generation. Many top managers have been conditioned by traditional management values. Perhaps their time frame and frame of reference makes them experience the fragmentation and the loss of cohesion more acutely than managers who are just starting out. Project managers work in an 'old' sector. The idea that the world could be changed (improved) was strong in the 1960s and 1970s, when – not purely by chance – co-operative development aid started to get going. The spirit of this era still haunts the Christian private development organizations in our second case. Possibly, the loss of this world vision is encroaching – consciously or otherwise – on the sense of purpose and commitment in relation to the work and its operational domain. However, if we compare both groups of managers and their ways of coping with the problems the management paradox creates, it seems difficult to maintain that top managers – the so-called 'winners' – do much better than the 'losers' from the ideological development aid organizations.

NOTES

1. That the introduction of instruments like the telescope and the computer had a huge impact on logic and the way of thinking according to us does not mean that we look upon the relation between culture and technology in general as a deterministic one. (New) technology has always had a strong influence, but the exact direction and the extent of the impact depend on a lot of other (environmental) factors.
2. This question is closely related to the first one. If we see culture as a fundamental human capacity to attribute meaning, we may also consider its function as a survival strategy. The reduction of uncertainty and the establishment of social ties may be viewed as such.

REFERENCES

Alvesson, Mats and Hugh Willmott (1996), *Making Sense of Management, a Critical Introduction*, London: Sage.

Arendt, Hannah (1958), *The Human Condition*, Chicago: University of Chicago Press.

Bass, B. (1985), *Leadership and Performance beyond Expectations*, New York: Free Press.

Bauman, Zygmunt (1995), *Life in Fragments: Essays in Postmodern Morality*, Oxford: Blackwell.

Beck, Ulrich (1992), *Risk Society: Towards a New Modernity*, London: Sage.

Beck, Ulrich, Anthony Giddens and Scott Lash (1994), *Reflexive Modernization: Politics, Tradition and Aesthetics in the Modern Social Order*, Cambridge: Polity Press.

Block, P. (1990), *The Empowered Manager*, San Francisco: Jossey-Bass.

Bos, René ten (2000), 'Fashion and utopia in management thinking', PhD thesis, Tilburg University.

Casey, Catherine (1995), *Work, Self and Society*, London: Routledge.

Deal, T. and A. Kennedy (1982), *Organizational Cultures*, Reading: Addison-Wesley.

Etzioni, Amitai ([1961] 1975), *A Comparative Analysis of Complex Organizations: On Power, Involvement, and their Correlates*, New York and London: Free Press.

Giddens, Anthony (1991), *Modernity and Self-identity: Self and Society in the Late Modern Age*, Cambridge: Polity Press.

Goffman, E. (1959), *The Presentation of the Self in Everyday Life*, New York: Doubleday.

Habermas, Jürgen (1981), *Theorie des Kommunikativen Handels*, Frankfurt am Main: Suhrkamp.

Hope, V. and J. Hendry (1996), 'Corporate cultural change, is it relevant for the organization of the 1990's?', *Human Resource Management Journal*, **5** (4), 61–73.

Kamsteeg, Frans H. (1998), *Prophetic Pentecostalism in Chile: A Case Study on Religion and Development Policy*, Lanham, MD, and London: Scarecrow Press.

Kluckhohn, F., F. Strodtbeck and J. Roberts (1961), *Variations in Value Orientations*, Evanston, Ill.: Row and Peterson.

Knights, D. and H. Willmott (1999), *Management Lives*, London: Sage.

Koot, Willem and Ida Sabelis (2002), *Surviving at the Top: Top Managers in Times of Complexity*, Amsterdam: Rozenberg Publishers.

Koot, W., I. Sabelis and S. Ybema (1996), *Contradictions in Context*, Amsterdam: VU University Press.

Kunda, Gideon (1992), *Engineering Culture: Control and Commitment in a High-Tech Corporation*, Philadelphia: Temple University Press.

Lazarus, R. and S. Folkman (1984), *Stress: Appraisal and Coping*, New York: Springer.

Maffesoli, Michel ([1988] 1996), *The Time of the Tribes: The Decline of Individualism in Mass Society*, London etc.: Sage.

Merry, U. (1995), *Coping with Uncertainty: Insights from the New Sciences of Chaos, Self-organization and Complexity*, Westport, CT: Praeger.

Mintzberg, Henry (1979), *The Structuring of Organizations: A Synthesis of the Research*, Englewood Cliffs, NJ: Prentice-Hall.

NWO (Netherlands Organization for Scientific Research) (2000), *Programma Sociale Cohesie* [*Social Cohesion Programme*], The Hague: NWO.

Pascale, R. (1990), *Managing on the Edge: How Succesfull Companies Use Conflict to Stay Ahead*, New York: Simon and Schuster.

Peters, Tom and R. H. Waterman (1980), *In Search of Excellence: Lessons from America's Best-run Companies*, New York: Harper and Row.

Schein, E. H. (1985), *Organizational Culture and Leadership*, San Francisco: Jossey-Bass.

Senge, P. (1990), *The Fifth Discipline: The Art and Practice of the Learning Organization*, New York: Double Day.

Smets, P., H. Wels and J. van Loon (1999), *Trust and Cooperation: Symbolic Exchange and Moral Economies in an Age of Cultural Differentiation*, Amsterdam: Uitgeverij 't Spinhuis.

Stacey, R. (1996), *Complexity and Creativity in Organizations*, San Francisco: Berrett-Koehler.

Van der Loo, Hans and W. van Reyen ([1993] 1997), *Paradoxen van modernisering*, Bussum: Coutinho.

Watson, Tony and Pauline Harris (1999), *The Emergent Manager*, London: Sage.

Wellins, S., W. Byham and J. Wilson (1991), *Empowered Teams*, San Francisco: Jossey-Bass.

7. The Discursive Constructions of Performing Professionals

Tom Keenoy

THE CONTEXT

In the context of the ineluctable socio-economic re-engineering which accompanies 'globalization', it seems none can escape the myriad forces reshaping the experience of work. While much attention has been paid to the politico-economic consequences of globalization (Giddens 2000; Ohmae 1994; Shaw 2000), in the longer term, the local–global cultural consequences may prove more fundamental. In this respect it is no accident that, since 1980, the language of anthropology has been progressively expropriated by those seeking to effect organizational change. Management policies and practices are now littered with references to culture, symbols, myths, values and artefacts (Casey 1995; Hannerz 1993). Lifted out of history and stripped of their original and nuanced complexities, these concepts have been employed as ephemeral tools in the never-ending quest for the 'one best way' to re-engage and re-motivate employees. This is one of the reasons why language has become a key resource in promoting, effecting and sustaining organizational change (Chouliaraki and Fairclough 1999; Grant, Keenoy and Oswick 1998; Keenoy 1997; Mumby and Clair 1997).

Analytically, the debate about organizational culture is articulated around the question, identified by Smircich (1983; see also Alvesson 2002), of whether such 'cultures' are the naturalistic outcome of participant interaction or whether they can be actively managed and manipulated. The central issue addressed in this chapter is the extent to which such detemporalized managerialist attempts to re-profile what might more accurately be called organizational 'ideo-cultures'[1] have been effective. Such initiatives are not merely designed to transform the organizational wallpaper but, as Fairclough (1999) intimates, the intention is to 'change hearts and minds' to secure 'cultural governance'.

Using discursive data from a research project which explores the changing nature of academic work in Britain, this chapter attempts to illustrate this debate and suggest an alternative way forward. In terms of their occupational discourse as professional workers, university academics are accustomed to relative autonomy in teaching and research; and they are generally scrupulous in their rhetorical defence of a professional identity historically rooted in the iconography of academic freedom. This term – while deeply ambiguous – carries a complex of organizational expectations and holds particular significance for the notion of managerial control. To put the matter sharply, the discursive presumption is that academics themselves should decide what should be taught and researched, and that they retain discretion (if not control) over how such work activities should be conducted. In such a context, academics ought to be better equipped than most to defend themselves against the more turbulent and disruptive consequences of those global waves.

In the account which follows, the conceptual-theoretic assumptions guiding the analysis are grounded in social constructivism (Gergen 2000); and, methodologically, the focus is on the discursive construction(s) employed to make sense of the new 'cultural' expectations which now confront British academics. Empirically, this involves a focus on organizational texts and the linguistic usage employed by academics to account for their work lives. This means that while the analytic focus is on the process of 'reality-construction', the structural-cultural context of 'text production' is also acknowledged (Fairclough 1995).

THE EMPIRICAL ILLUSIONS

At the level of organization, one expression of those global forces has been the plethora of managerial attempts to 'transform' employee meanings and expectations through organizational change programmes (Keenoy and Anthony 1992; Martin 2001). And, whatever the experienced facticity of the discursive promise of HRM (Keenoy 1999), there seems little doubt the lifeworlds of contemporary employees and the nature of work organization bear little resemblance to those of the 1970s. While talk of transformation may be premature, the contemporary experience of work is now qualitatively different (Brown 1997; Casey 1995; Gallie et al. 1998; Noon and Blyton 1997).

But the evidence seems paradoxical. On the one hand, there is much to suggest the language and imagery of HRM is not mere rhetoric. Throughout the 1980s and 1990s the now ubiquitous discourses of flexibility, customer care and total quality, and their associated mechanisms, have generated a range of 'employee-friendly' polices and practices designed to increase autonomy, extend participation and promote greater employee-responsibility (for example, Frenkel et al. 1999); all of which ought to make us all feel better about ourselves, our work and our work organizations. In contrast – and in parallel – there is also much to suggest that not only has employment security disappeared into history but also that the experience of many recent HR-type innovations appears to amount to little more than work intensification and – in what is undoubtedly an older discourse – an apparent increase in alienation (for example, Taylor and Bain 1999). While some have been 'incentivized' to become the eponymous committed employee, there also appear to be many for whom permanent insecurity may be the driving motivation to 'perform'; all of which ought to make us all feel worse about ourselves, our work and our work organizations.

In short, we appear to be in a situation where increases in employee freedom and responsibility are coterminus with increases in managerial control, insecurity and worker exploitation – sometimes in the same organization at the same time (Legge 1995; Watson 1994). In the rational, depoliticized, dualistic world of managerial texts it is usually assumed that the simultaneous pursuit of such apparently incompatible control strategies is unwise and, more importantly, likely to be ineffective. However, it will be suggested that there is no necessary contradiction in this development; indeed, such apparently contradictory practices have long characterized management practice.

WHY UNIVERSITY ACADEMICS?

In terms of images, it might be thought the reflections of university academics, traditionally cloistered beyond the reaches of anything so profane as a labour market, have little to tell us about the contemporary experience of work. However, one of the reasons for researching academics is because the management strategies now employed in British universities appear to draw more on the model of industrial bureaucracy than the 'people-centred' models associated with the knowledge industries. Paradoxically, not so long ago (Kanter 1988) it was being argued that industry and commerce needed to learn from the universities; that the trust-relations, the autonomy, the 'inner-directedness' and the patterns of self-organization found among academics was what organizations needed to for success in an increasingly competitive globalising world. In short, success depended on becoming more like universities.

Unfortunately, successive British governments since the late 1980s were unaware of this insight and opted instead for a solution which embraced the discourse(s) of public accountability, cost-effectiveness and value for money (Power 1997), and led to what has come to be known as the 'new public management'. (For an excellent summary overview of the arguments surrounding new public management, see Reed 2002.) This trajectory of change, which has been continued and, to some degree, intensified with the election of Mr Blair's New Labour administration in 1995, privileges productionist values which appear to have been institutionalized through the imposition of a series of auditable output criteria relating to, primarily, teaching quality and research activity. In effect, in the legitimatory context of a dramatic expansion in student numbers in higher education, the traditional high trust managerial strategy which permitted academics considerable discretion in their choice of work activities is, increasingly, being weakened by a raft of externally generated performance indicators which find their origin in the values of management accountancy. While these retro-controls are a long way short of the Fordist assembly line, the apparent increasing commodification of academic work and the alleged McDonaldization of higher education have not gone unremarked (Parker and Jary 1995; Prichard and Willmott 1997; Willmott 1995). The focus here is on academics' work as 'researchers' and, in particular, their responses to what is known as 'the RAE' (Research Assessment Exercise). This is the managerial mechanism through which research activity and research performance are periodically audited and evaluated in British universities.[2]

The Research Assessment Exercise (RAE)

To date, there have been four such audits. The first, in 1989/90 was a low-key affair; the next three, in 1992, 1996 and 2001 have been far more robust and, bureaucratically, complex and demanding. In the briefest of summaries, the exercise involves university departments submitting a rigidly standardized return reporting staff research output ('measured' by four publications) and detailing their research income and activities over the assessment period. The 'return' – as it is known – also requires departments to include more qualitative statements relating to research groupings, 'research culture', research strategies and scholarly status. Performance is evaluated by 'subject panels' – comprised primarily of respected academics – established for each academic specialism. The 'returns' are the basis for a comparative evaluation of the relative performance of all departments submitting a return within each subject area with departments being ranked on a seven-point scale (from 1 to 5*). Particularly in the older universities, the RAE is, perhaps, the major mechanism which has driven organizational change since 1989. However, it is not mandatory: individual institutions decide which departments will submit a 'return'. Failure to submit means a department is not eligible to share in the significant research funds allocated through central funding.

For many departments, this single figure, the 'RAE rating', calculated once every five years, is an extremely important signifier. Operationally, the rating determines a significant proportion of funding but, iconographically, the numeral is a vital sign. It indicates progress, status, achievement, 'quality', influence and, crucially, whether a department is going 'up or down'. A 'good rating' serves as a marketing emblem; equally, a 'poor rating' is thought to be a disincentive in attracting better students and staff. Although the RAE has had a differential impact, organizationally, there can be little doubt the introduction and progressive incorporation of this 'modernizing' initiative has had a fundamental impact on the cultural context of university organization and the (work)lives of academics. The empirical focus of this chapter is to provide an illustrative account of how the RAE process has ineluctably affected an array of mutually implicated consequences which have shaped, interacted with and/or impacted on organizational structuring and the experience of work.

METHODOLOGICAL PRAGMATISM

One of the possible reasons for the absence of extensive studies of academics' work experience may well be that 'studying oneself' involves some extremely troublesome methodological issues. Indeed, it is difficult to evade the implication of one academic who, on being told about the research, laughed and said: 'That's impossible; it's like *pretending* to be a fish on the outside of the goldfish bowl talking about what it's like to be on the inside' (WS).[3]

And the apparent problem goes deeper than potential 'observer bias' for I am inextricably implicated in the practices and irreversibly embedded in the values of *our/my* new 'performance culture'. Moreover, my 'pre-judicial' experience is layered in the managerial policies and professional practices of both the pre- and post-RAE 'linguistic regimes' (Cameron 2000). Of course, this is not a new but merely a more self-evident instance of the ancient 'problem' of social values in social scientific research.

Hence, in order to discipline the intrinsic (and inescapable) 'contamination' resulting from my co-construction of the informants perceptions, it was decided to employ two methods of data collection. First, from about a year before the fieldwork commenced, conversations with academics in a variety of naturalistic settings relating to the experience of contemporary academic work were recorded in a research diary. The collection of such episodes is not without problems (Coffey and Atkinson 1996; Silverman 1999). Unlike the material produced through the interviews, it provides less mediated illustrative examples of academics' everyday expressions and conventional discursive usage. While collected opportunistically, this data provides *some* re-presentation of the 'temper' of the community of practice (Wenger 1998) being explored. Secondly, to date, 33 interviews have been conducted with samples of academics in a number of departments of management. Respondents were selected to approximate a representative balance between academic specialisms, age, gender and status.

On the speculative assumption that high-performing departments are likely to have embraced the ideo-culture of the RAE more enthusiastically, all the departments selected had an RAE rating of 4 or better (Trowler 2001). The expectation is that such departments might provide evidence of a wider range of change and a more diverse set of managerialist policies and practices designed to 'improve' their RAE ratings. In this respect, in terms of case-study methodology, the departments should be seen as 'elite' cases deliberately selected to illustrate a predicted phenomena (Eisenhardt 1989; Yin 1989). In other words, they tell us what they tell us.

The recorded interviews were non-directive and – in so far as it is achievable in an artificial setting – as 'conversational' as possible. In requesting their co-operation, respondents were only told: 'The project is a study of the changing character of academic work in the light of the various changes which the Higher Education sector has been subject to in recent years'. And, in the interview, after reassurances about confidentiality, the research was contextualized as 'an old-fashioned sociology of work project'; the RAE as such was deliberately decentred by saying: 'I want to talk about how academic work has changed while you've been an academic'. Having secured a short career biography, the exchange was managed through a series of open-ended simplistic questions about the nature of work, work organization, factors effecting changes to work and the effect (rather than 'impact') of the RAE on the interviewee and others. It ended with the traditional questions about the three things respondents liked and disliked about their work. In between times, respondents were encouraged to follow whatever line of thought came up. Interviews varied in length between approximately one and two hours; there was very considerable variation in both the framing of answers and the issues covered; not all were 'successful'; not all included each and every question on the schedule; and while most were held in the respondent's room, some took place in cafeterias.

Overall, the intent was to collect accounts of how work has changed throughout each individual's career in the expectation that this would provide discursive reconstructions of how the experience of work has developed through time. Despite its significance in the linguistic and policy regimes in contemporary universities, the RAE, as such, was not privileged.

THE EMERGENT CULTURAL-STRUCTURING

At the level of work organization, the surface daily routines appear to have changed little. For most informants work still comprises a variable mix of teaching, research and administration. However, the character and, to a lesser degree, the content of their work-activity has changed; the work they privilege and prioritize has changed; and there appear to have been some subtle but significant changes in work-orientation. Perhaps unsurprisingly, it seems to be the work organization of senior staff which has changed most. In general, all the 'front-line' academics I talked with confirmed Reed's analysis – which focused on 'academic-managers':

> Overall, our research data indicated a generalized perception of increasing managerial *and bureaucratic* control with declining levels of trust in and discretion afforded to professional academic mores and practices. While this generalized perception of 'advancing managerialism' was expressed more strongly by non-managerial academic staff, even amongst our manager-academics there was a very clear view that new managerialism – as a 'theory of practice' – had deeply and extensively permeated UK universities' organizational cultures and practices over the last decade or so.
>
> (Reed 2002, p. 191; italics in original)

In terms of (departmental) organizational re-structuring, his findings are reflected in the fact that all the departments studied now have someone specifically designated to co-ordinate research strategy and oversee the periodic production of the RAE return. Similarly, all also have a Research Committee which is responsible for promoting research; collating information about research funding and co-ordinating bids for research grants; and, generally, developing the 'research culture'. Such departmental committees are invariably linked into a structure of higher-level institutional bodies which monitor and oversee the work of those below them. For some universities, strategically, research income is vital and some university-level research bodies have extensive powers: 'We've been told that anyone with less than a 3 [RAE rating] is going to be "visited" to improve their performance' (MS). While few informants made specific reference to such bodies, it was clear they were aware of what was expected of them: 'I've already got my 4 pieces [that is publications] for next time' (WJ).

The progressive importance of the 'research culture' as a behavioural reference point is perhaps reflected in an apparent shift of emphasis in work roles. When asked about their 'work', most of the interviewees spent far more time talking about their research activities than anything else; indeed, generally, interviewer prompting was required to get them to detail their teaching 'work' and what that involves. While there was no suggestion that teaching is not treated as a significant work commitment, it seems as if teaching activities are something to be completed before individuals 'get back to' their 'real work' – research and publishing. Such attitudes are not new and, while they may now be more widespread, it should be emphasized that this privileging of research over teaching was by no means uniform. Senior staff also talked more about their administrative roles than teaching; and, whilst this too may not be new in itself, it seems clear there has been a very significant increase in the administrative demands (on all staff) consequent on the various externally imposed audit mechanisms. Apart from an acute awareness of the 'need to publish', what also seems 'new' is that many now engage in a far wider range of research-related activities: there is frequent mention of research seminars, conferences, workshops and day-schools as 'normal events' within the 'daily routine' and such events are often directly related to what one called 'publishing opportunities' (MJ). Structurally, this is also reflected in what appears to be an increase in the number of dedicated 'research units' either within or attached to academic departments – such 'units' are believed to 'count' in the RAE decision-making process. Academics within such units tend to have minimal teaching responsibilities.

These developments suggest that an increasing division of labour within the academic lifeworld has been one effect of the new managerialist emphases on performativity. For example, in one of the departments studied, there are a notable number of lecturers employed on short-term contracts and others on 'teaching-only' contracts (such employees do not 'count' in the RAE and may help to reduce the teaching loads of 'research-active' staff) together with a number of more specialized research-orientated posts which carry much lower teaching loads. While such initiatives are not uncommon elsewhere, despite 12 years of the RAE linguistic regime, they are by no means the 'norm'. But it would be unwise to attribute such developments solely to the impact of the RAE. While it seems clear there *has* been an increasing differentiation in the activity-mix of academics' work content, this is occurring in the wider context of an overall expansion of higher education and academic research. In this respect, there seems to be increased differentiation both between and within universities, and expansion, of itself, has, perhaps necessarily, been accompanied by greater rationalization in work organization. One feature of this has been increasing size which – independently of the periodic teaching and research audits – permits a greater degree of specialization. Against that, it should be remembered the additional funding from a 'good RAE rating' means 'better' departments are in a position to keep individual teaching loads lower than elsewhere. In this respect, the RAE – as a managerial mechanism – induces a 'natural' (and some might say, 'market-driven') increase in specialization. This means research-orientated individuals are simultaneously both attracted to and rewarded by higher-rated departments. Unsurprisingly, 'better-published' research-orientated individuals gravitate toward departments which promise lower teaching and more time for research as well as, in many cases, better promotion prospects. This was not always the case and, as these latter remarks imply, the increasing division of labour has not necessarily been resisted nor regarded as 'de-professionalizing' – at least, that is the predominant impression gained from the interviews.

Underpinning these developments in the cultural-structuring of contemporary academic worklife are the pervasive ideo-cultural values of 'visible performativity' (Deem 1998) on both individuals and institutions. For individuals, this comes primarily in the form of periodic (and formal) performance appraisals and routine if intermittent informal scrutiny of what is often referred to as 'RAEable research activity' (MS). The objective of both is to ensure that staff have or will have 'four good pieces' – a euphemism for four articles in highly rated academic journals – to submit for the next RAE. Indeed, the impression gained is that 'minimal acceptable performance' might be reducible to this single criterion and individuals appear to be subject to a range of persistent stimuli to ensure this. For example, performance appraisals might focus on planning 'research outputs'; teaching might be concentrated in one term to ensure additional time for writing or research; senior staff read draft articles and advise both on the content and on where and how to get published. Similarly, they advise on sources of research funding and how to draft effective funding applications; and the presentation of conference papers – which may lead to publications – is encouraged (and funded). As one respondent put it: 'The walls are constantly echoing research' (WJ). While some experience such activity as pressure, for most it appears as simply 'the way things are' and – at least in terms of the responses – is not perceived as untoward or problematic. Inevitably, however, some individuals – mainly among the longer-serving staff – feel marginalized by these organizational emphases.

Bearing in mind that generalization may be no more than an attempt to inject analytic security into an intrinsically ambiguous social reality, the emergent cultural-structuring appears to be constructing a progressive work intensification in combination with a re-definition of what is deemed to be *desirable* academic research activity for, in general, the value of short-term measurable outputs appears to be privileged over longer-term more nebulous achievement criteria. 'Everybody says you shouldn't write a book for all sorts of reasons' (MJ). But, to re-iterate, it would be premature to conclude there is convincing evidence of 'de-professionalization'. Whatever is happening appears more akin to a process of 're-professionalization'.

MAKING SENSE OF PERFORMATIVITY (PART ONE)

Despite the apparent straitjacket of performativity criteria, individual academics make sense of these new imperatives in qualitatively different ways and display a variety of psycho-social and task-related coping strategies to negotiate this 'new reality'.

At one extreme are those who continue to reject the 'new managerialst' agenda or, in some cases, consciously choose to ignore the new performance criteria. (This is certainly not to imply they are not 'performing' in a managerially acceptable fashion.) One individual (MJ) claimed not to know which of his publications had been entered in the recent RAE submission, 'I just see the frenzy, I don't get involved', and, referring to the impact of the RAE on the culture of a department in which he had previously worked, he described it as: 'extremely – very – competitive; [it's] a horrible culture to be in [and] the RAE nurtures an awful lot of that … I'm dead against that'. Similarly, with respect to the teaching audit process (known as 'the QAA') he insisted: 'it hasn't changed what I do' for, despite the QAA audit criteria, 'what I believe makes good teaching is not measurable … I will pay lip service to them [the QAA criteria] ... I just totally disagree with everything being prescriptive ... it's totally alien [to what a university stands for]'. He claimed not to be under any pressure to conform (while also conceding his nonconformity 'probably explains why I'm still a lecturer'). In similar fashion, another (WJ) who also continued to privilege teaching over research values, insisted 'there should be room for lecturers who lecture' and thought that the QAA 'makes less difference on the ground than one would think'; that it has caused 'more time and more time in meetings talking about things' and 'it's contributed to everyone's resentment'. On the impact of the RAE, she claimed it had hardly affected her at all since she was no longer research-active. 'I wouldn't say I never have ideas but … I would have to stop doing one of these other things [teaching and a heavy administrative load] or stop sleeping altogether.' Nor did she feel in any way devalued because 'I can't convince myself that what I'm doing here is not a [valuable contribution]'. She reckoned she worked around 65 hours a week and that 'I really object to the way in which research has come to dominate academic life'. More generally, she observed: 'I can't help feeling that … well, who gets excited about these things? … most of it is so bloody pathetic ... my values vary so considerably … I just get on with what I think I should be doing [even though] they think I'm not good at the things that matter'; and 'it's suited some people to think … we can make [the university] better by shoving it up these lists.'

Such resilient sense-making was untypical (both had entered the profession in pre-RAE times). However, even among those who have complied with the new expectations, the legitimacy of such 'traditional' readings of the impact of the new performativity values on academic worklife is often clearly acknowledged. That said, at the other extreme, there are individuals who display considerable enthusiasm for making performance measures 'visible' and actively measuring outputs. One individual (MS) – who had reached a point where he thought 'am I going to do this [teaching] for the rest of my career?' – experienced the RAE as a new stimulus: 'research was a new personal direction for me'. He was unsympathetic to those who thought academic work had become more pressured – 'if you get stressed in this job, there's something wrong with you' – and, emphasizing the fact that universities run on public money, pointed out that 'we're just being asked to do our jobs' and that 'we have to be accountable'. But, faintly echoing those 'older' values, he also complained about younger staff who are so research-orientated 'they don't want to teach' (by which he meant they do not engage in the 'value' of teaching as a core academic responsibility). More generally – and with no hint of inconsistency – he also claimed to be 'an old cynic' and that the RAE may be there to 'catch the 5 per cent of rogue academics' who have always under-performed. While few framed the issues as graphically as this individual, the discourse of 'being-accountable-for-public-money' informed a significant number of respondents' self-conscious justifications for conforming to (or not resisting) the new cultural structuring.

All the individuals quoted thus far became professional academics prior to the introduction of the RAE. However, among those who have started their careers since the RAE became a key element in the ideo-culture there appears to be an equally diverse range of responses. The visibility of the new linguistic regime was highlighted by one individual (MJ) who, reflecting upon his recent induction day, said:

> the thing that stuck me about the [talk] on teaching [was that it] purely focused on the QAA and on the empirical measurement of our teaching [while] the research talk placed great emphasis on the RAE ... and how significant this was to the Department and, 'yes, it's an imperfect system, but it's a competition and we want to win' and 'yes, it's the bottom line' … [these induction talks] were indicative of the priorities.

All this was an 'absolutely a new emphasis' compared with the approach adopted by his previous (lower-rated) university and, moreover, he suggested, 'when they say research, they mean publications … you can't spend 10 years reading and researching [before you publish].' Nevertheless, he appeared untroubled by the RAE: 'apart from just filling in endless forms and the various micro controls they impose on you … the RAE is part of the research culture here … in that way it's affected me psychologically, culturally … I've put in some publications but am not quite sure whether they will go in'. Asked to elaborate on what he meant by 'micro-controls', he said: 'the form filling was one thing ... the talk given by the Research Director, the interview with the Head of Department, the [job] interview … what my mentor says … the talks you have with colleagues.'

Such comment suggests how the ideo-culture is transmitted through the combination of routine linguistic framing and institutional reinforcement; it adds some substance to the idea of 'walls echoing research'. This remark came from an individual (WJ) at another institution who also seemed unfamiliar with what might be expected of her. In contrast to previous respondent, she continually referred back to the RAE as a source of pressure. Asked what was her work, she replied: 'I would put research first simply because of the pressure put onto me by the Department followed by teaching and administration ... It wasn't what I expected ... I didn't realize that it was so enforced.' She 'was expecting research to be more collaborative [since] to get new stuff going is quite a daunting thought [and] I don't like working on my own'. And, explaining where the pressure came from, she said: 'Well … basically pressure from the RAE … but I think the Department doubles that up … we're expected to submit 6 [articles] so they can submit 4' (this 'fact' was not mentioned by any of the other interviewees at this department); and other pressures came from 'a lot of stuff sent round by email [and] then there is the … I think she's called the Research Coordinator … and just the ambience of the place'. Such concern also seemed to be underpinned by uncertainty: 'I don't know how the whole thing works [and] I think the pressure's on even in terms of moving jobs ... so you can't even think in terms of wanting to relax … You're just left there.' However, although not included in the recent RAE submission, she had sent two draft articles to journals knowing they would be rejected because 'I thought I'd save them for the next RAE'.

While others also displayed a perhaps surprising lack of knowledge about the RAE process, few *expressed* the same concerns about experienced pressure. One person (MJ) – who was co-editing a book – appeared unaware that such scholarly activity was not highly valued in terms of the RAE Panel's prevailing criteria. 'Well, if he'd [the Head of Department] told me that, I'd have told him "I'll do 4 articles as well".' As for the RAE itself: 'I don't like to say this, but I quite like it … I love writing … it's what we're supposed to be doing anyway.' Another (MJ) seemed equally relaxed: 'I mean the RAE was coming up … so there was some pressure to have something somewhere ... not an explicit pressure … of course, the implicit pressure is there, from probation [new lecturers spend three years 'on probation' before their appointment is confirmed] and also from the general expectations which are raised … I won't say it's a pressure which causes me nightmares … it's a thing I would do anyway … but [it's] not: "I now *need* this publication, *this* year" … that's not how I perceive what's going on.' And another (MJ) also on probation, said: 'It [the RAE] doesn't worry me too much … I feel reasonably pressured ... I feel more pressure to upgrade' (that is, to get published in higher grade journals).

Others displayed a clear strategic awareness of how to 'play the publications game'. One (MJ) detailed the issues around choosing 'suitable' journals in which to place work. 'We had an article which was sort of political economy but sort of public policy [and] without the RAE I think I would have agreed easy ... you do agonize about "is it [the journal] RAEable?"' Another (MJ), 'deliberately targeted [top journals] in order to get the best referees' comments … I really did not expect them to be accepted … indeed … I would have been astonished'. And the extent to which RAE-related performance is linked to career progression was illustrated by the comments of someone who left his previous post because he was 'under pressure to alter my research … [this was] not wise for my own career and I didn't think it was necessary'. There was 'an issue about whether to include me in the RAE [and this was] the final straw ... I saw that as professional death' (MJ). Similar, if more subtle, pressure to conform to 'the RAE format' was expressed by others. Some felt pressured to join 'research groups' while others were 'counselled' about developing 'more suitable' research areas; few would contemplate longitudinal research; and one claimed that difficulty with finding a publication which would take her work meant that: 'I decided I was better off abandoning the qualitative idea' (WJ). At more senior levels the issues are slightly different. 'As long as stuff is being published you're not under pressure ... it's like the goal-scorer [in football]. I always like to be in a comfortable position with respect to the RAE ... what I've been doing is having 4 or 6 or 8 [articles to choose from] .... Some people play a different game – they just produce 4 in highly rated journals' (MS).

One theme which is beginning to come through the data is an apparent instrumentality – or, perhaps better, a 'situated pragmatism' – both on the part of individuals and departments. The apparent 'need' for such tactics is illustrated by one of the senior respondents (WS) who, clearly concerned to protect her less experienced staff and jealous to maintain what she regarded as academic autonomy, explained that:

> We play the numbers game [and] we have a portfolio approach to our CVs ... we make sure we get publications in journals that are rated – those are where we get the RAE points – as well as putting in papers to encourage some newer journals ... we try to get some kind of balance ... we try to be broad and include all sorts of perspectives, for example, if someone got interested in the political science perspective, we'd say 'fine' [but] we look at [the overall trajectory of] publications. Each individual has a writing plan ... it's difficult . . . everybody has a writing plan which they are working on ... and we look at what's happening ...

Such cautious euphemism might well be a description of a carefully calculated strategy to ensure everyone maximizes their 'RAEability' (while simultaneously promoting individual career paths and guarding academic autonomy). Similarly, another senior person, commenting on the practice of publishing multi-authored journal articles (a practice which, on at least some occasions, might reflect a pragmatic concern to ensure individuals have a sufficient number of articles to submit for the RAE) insisted: 'We *do* actually jointly write things ... the whole team is involved ... and that's part of the fun' (MS).

This situated pragmatism is also reflected in the emergence of a more coherent and intermittently operative labour market for academics which is clearly related not only to the changing character of the higher education sector in itself but, more specifically, also to the arrival of the RAE. 'If you set up a particular set of incentives you can't be surprised if people respond to them' (WJ). In the formal interviews this feature was often tastefully marginalized if, invariably, tacitly acknowledged. 'Arguably the RAE gives you some more bargaining power in terms of salary ... I'm aware of it's potential use – the other day a colleague ... he said it's interesting [that] we never bargained with anyone before … hiring takes longer now … because people are bargaining' (MJ). 'Transfer market? Is there a benefit there? "My CV's bigger than yours?" ... there might be an individual benefit … yes, I suppose there are some people who've benefited but they're probably quite good anyway' (MS).

The apparent studied indifference of such remarks is deceptive for it belies both the widespread awareness of what has come to be called 'the transfer market' and the competitive – even 'enterprising' – character of less guarded comments made in more naturalistic settings: 'I am a 5 star performer' (MS); 'I need to get some more stuff [publications] under my belt before I can hit the chair circuit' (MS); 'No one bothered [about us] before … now I'm valuable once every 5 years; why shouldn't we look after ourselves?' (MJ); 'He's just building his CV' (MJ); and 'Of course you have to look after yourself' (WS). While some have always pursued the careerist implications carried by such remarks, in parallel, there are also a significant minority who regard what has been happening as undesirable: 'No one thinks about loyalty or responsibility any more' (MS); 'We're demeaning ourselves … these days … any 25-year-old with 4 anonymous publications can get himself a chair in a 'new' university' (MJ). Despite the self-evident hyperbole of this last comment, following what was seen as excessive 'transfer activity' before the 1996 RAE, the authorities changed the procedural rules for the 2001 RAE in an attempt to reduce the incentive for departments to 'buy themselves a rating'. There is little evidence this was effective. Indeed, the reverse might be the case; for at least a minority of departments appear to have become even more strategic – offering a range of posts with low teaching responsibilities – and some now talk in managerialist fashion of exploring ways to 'anchor staff' (MS) in departments. As ever, it is more complex than this; but few would disagree there has been an sharpening of individualistic attitudes among academics. (Of course, the RAE is by no means the sole cause.)

MAKING SENSE OF PERFORMATIVITY (PART TWO)

Returning to the issues which framed this chapter, it now seems possible to offer some provisional answers to the questions about the nature of organizational culture(s) in the context of new forms of managerial control.

Following the implementation of the 'new managerialist' performativity mechanisms, discursively, it seems three inter-referential (Eco 2000) linguistic regimes presently inhabit elite departments in British universities. First there is the *expressed ideo-culture* of audit and performativity which is the 'official' or formal organising culture. Despite qualified reluctance, it is promulgated and promoted by academic-managers and – with sometimes greater reservations – most academics now appear to accept it even if they do not *expressly* endorse it with any enthusiasm. While widespread behavioural compliance is no indicator of positive engagement, it does seem to be accepted as a framework of legitimate expectations. As one person put it: 'Yes, it's [the RAE] now part of the wallpaper' (MS).

Secondly, there is the *expressed 'traditional' culture* which idealizes the historic icons of academic freedom and dispassionate reflection and informs the (self-)image of the academic as teacher, knowledge-pursuer and knowledge-creator. 'Y'know, I still think of myself as a teacher' (MS) and 'But, we can still do "good work"' (MJ) are comments which both came from 'high performing' academics. In this respect, it would be a serious misjudgement to regard this iconographic traditional culture as intrinsically antipathetic to the ideo-culture. While a small number of respondents could be described as 'anti-RAE', their influence is marginal and, apparently, declining. Habituation and the slow but relentless process of staff turnover are working over any residual resistance. However – and this is no paradox – the historic core values they express littered the discursive surface of *every* interview. (This holds significant implications for the 'emergent' academic 'identity', for the process of re-professionalization and for the terminally perplexing issue of what people 'really' believe.)

And, thirdly, there is the *expressed 'coping-culture'* which was characterized above as one of situated pragmatism. In common with employees everywhere, academics have to find ways of making sense of the projected abstracted realities which form the basis of managerial ideo-cultures. They see themselves as having 'no choice' but to accommodate the various initiatives and have 'translated' these contextual realities into acceptable forms of work organization and performance. As noted above, academic discourse indicates that a rich diversity of individual and institutional responses have been devised which, simultaneously, appear to meet, circumvent, manipulate and, on occasion, subvert the ambiguous and cumbersome mechanics of the accountability criteria. As someone who held responsibility for implementing the teaching standards remarked: 'Teaching quality? What's that? I wish someone would tell *me*'(MS). At the same time, they seem to believe they can mitigate the impact of the new criteria which, generally, appear to be interpreted as conditioning rather than determining behaviour: 'Yes, of course, the RAE is nonsense ... it's journals which count not the content ... but we have to make it look good' (MS). Needless to say, there are now many more individual academics who are highly skilled in the semiotics of 'looking good'.

This is not to suggest academics do not do what is 'expected' of them; far from it. Indeed, if it is possible to venture such a judgement, most appear to be working longer hours and producing more 'outputs' than in pre-RAE times. Certainly, far more is expected of younger staff than was the case pre-1989. The point is that they 'appear' to perform in a visibly measurable and acceptable fashion while attempting to minimize the potential intrusiveness and – as they might see it – the potential destructiveness of the new performativity demands. More generally, in so far as the reported behavioural changes accommodate the formal accountability criteria of the 'new public management', we might even conclude that the government-initiated 'culture-change programme' has been 'efficient' (although, of course, it remains impossible to judge whether it has been 'effective').

With respect to the simultaneous increase in both employee freedoms and managerial controls, the situation appears ambiguous and open to several interpretations. At best it could be said that the 'shape' of the autonomy–control relationship has been reconfigured. While some 'traditional' cultural time-spaces seem to be closing down, new ones have been opened up (at least in elite departments). If individuals frame their work activity within the script of 'RAEability' it is possible to carve out more 'freedom' than existed before. This is particularly so for younger staff who develop a 'good research track record' which is likely to make them 'more marketable'; such a profile means they may be able to place themselves in a virtuous circle of decreasing teaching and expanded research time which, in turn, opens up the promotion ladder at an earlier age. However, such a time-space is highly conditional on meeting managerially specified targets and might require a reinterpretation of the 'traditional' notions of academic freedom and self-organization. In addition, as indicated in the discourse of the two individuals who had resisted the ideo-culture of accountability, for the time being, it also remains possible to continue to operate according to those older, less conditional values. Any judgement here is made all the more difficult because, as contexts change so to do the evaluative criteria we employ to characterize them. What might be the analytic currency of 'exchanged freedoms'?

One Final Possibility

This last question – to which there can be no satisfactory answer – might suggest we need to ask the question differently. It is not 'is academic work more controlled or more free' but – since we are exploring how constrained employees have responded to a process of externally imposed organizational change – 'how do academics (re)construct their "new" social reality?' From the interviews, it seems clear there is a complex variety of multiple, co-existing and sometimes apparently contradictory cultural norms which individuals draw on in making sense of and ordering their everyday experience. And, for the most part, they do not appear operate on the basis of fixed ideal-typical notions of 'culture' or 'values'. In conversation, they draw selectively on ideo-cultural, 'traditional' or coping-culture legitimations depending on the subject. In other words, different values (or evaluations) appear to be privileged in different contexts. For example, a number of respondents acknowledged the 'need' for universities to be 'publicly accountable' while simultaneously insisting that nothing should be permitted to limit 'academic freedom' and agreeing that RAE is 'not all bad' because it provides rewards for those who perform to the desired script. In the extreme, this is arguably indicative of inconsistency, confusion and incoherence in their perceptual frameworks.

Alternatively, we could accept such 'incoherence' as a normal and predictable feature of their differential engagement in the organization, their professional identities and their careers. Indeed, from a social constructivist perspective – which privileges the perspective of the constrained social actor – this is what one might expect to find. This perceptual instability and intrinsic ambiguity refract the apparent contradictions and multiple meanings which are invariably found in organizational lifeworlds (Gergen 2000).

So, in what sense is it meaningful to talk about 'the culture' of contemporary organizations? The academic social actors considered above appear to share so many aspects of their permeable location so why not also an organizational culture? However, as demonstrated above, this 'culture' is a multi-faceted phenomena which – when we go looking for it – presents itself as something different depending on the context. While it is tempting to resolve this apparent problem by fragmenting the 'data' and resorting to the dualistic solution of identifying 'counter-cultures' or 'sub-cultures', this misses the crucial point that social actors have no problem in routinely reconciling apparently contradictory evaluations which make sense of their social reality. Such a solution also maintains the analytic fiction that 'a culture' should be coherent and consistent.

A more realist(ic) solution is to acknowledge that any organizational 'culture' at a particular point in time will both refract and embrace a complex of alternative possibilities reflecting not only the socio-economic context of action but also the complex histories and present engagements of those who comprise 'the organization'. Hence, the various cultural orientations identified above can be seen as mutually implicated in each other and *all* those contradictory elements can be regarded as intrinsic to whatever it is we mean by the term 'organizational culture'. This 'culture' conditions not only what people *say* they do but also what they do: perhaps they achieve 'coherence' through necessary inconsistency? In this respect, culture is not unlike a hologram (Keenoy 1999) which – as one moves around it – presents us with a different aspect of its 'being'. All the facets of a hologram are inextricably part of the whole while, in turn, each narrow facet is representative of and contains the whole. It would seem that if we are to get closer to or even grasp what 'culture' is we need to be able to understand it as something which – like the hologram – is a fluid ever-changing phenomena which is constantly 'adjusted' or 're-invented' by social actors as they encounter, experience and address emerging contextual realities. In other words, intrinsically, culture 'is' an emergent process, not a fixed or fixable form.

NOTES

1. This term evades the 'culture problem' identified by Smircich (1983). It refers to managerially initiated and managerially driven conceptions of what constitutes appropriate organizational behaviour. 'Ideo' to indicate that such conceptions reflect managerial objectives, and 'culture' to indicate that such expectations are either implicit in 'new' workplace regimes and/or explicitly legitimized through 'new' norms and values.
2. An elaboration of how this auditing process works can be found at the Higher Education Funding Council for England (HEFCE) website: http://www.hefce.ac.uk/.
3. All remarks in double parentheses are direct quotations from conversations or interviews. The notation indicates gender (M or W) and status (S or J) of the informant. In British terminology, 'Senior' includes professors, readers, senior lecturers and their equivalent research posts; 'Junior' (J) refers to lecturers, teaching-only staff, tutorial assistants and their equivalent research posts. (This is merely a satisficing solution to the ambiguities which characterize almost any categorization process. For example, in Britain, there are many very experienced and highly knowledgeable lecturers for whom the designation 'Junior' might appear demeaning. My apologies. Similarly, these days, it is not unusual to find relatively inexperienced senior staff.)

REFERENCES

Alvesson, M. (2002), *Understanding Organizational Culture*, London: Sage.

Brown, R. K. (ed.) (1997), *The Changing Shape of Work*, London: Macmillan.

Cameron, D. (2000), *Good to Talk*, London: Sage.

Casey, C. (1995), *Work, Self and Society*, London: Routledge.

Chouliaraki L. and N. Fairclough (1999), *Discourse in Late Modernity: Rethinking Critical Discourse Analysis*, Edinburgh: Edinburgh University Press.

Coffey, A. and P. Atkinson, (1996), *Making Sense of Qualitative Data: Complementary Research Strategies*, London: Sage.

Deem, R. (1998), 'New managerialism in higher education: the management of performances and cultures in universities', *International Studies in the Sociology of Education*, **8** (1), 56–75.

Eco, U. (2000), *Kant and the Platypus*, London: Vintage.

Eisenhardt, K. M. (1989), 'Building theories from case study research', *Academy of Management Review*, **14** (4), 532–50.

Fairclough, N. (1995), *Critical Discourse Analysis*, London: Longman.

Frenkel, S., M. Korczynski, K. Shire and M. Tam (1999), *On the Front Line: Organization of Work in the Information Economy*, Ithaca, NY: Cornell University Press.

Gallie, D., M. White, Y. Cheng and M. Tomlinson (1998), *Restructuring the Employment Relationship*, Oxford: Clarendon Press.

Gergen, K. (2000), *An Invitation to Social Constructionism*, London: Sage.

Giddens, A. (2000), *Runaway World: How Globalisation is Reshaping Our Lives*, London: Profile.

Grant, D., T. Keenoy and C. Oswick (eds) (1998), *Discourse and Organisation*, London: Sage.

Hannerz, U. (1993), *Cultural Complexity*, New York: Columbia University Press.

Kanter, R. M. (1988), *The Change Masters*, London: Unwin Paperbacks.

Keenoy, T. (1997), review article: 'HRMism and the languages of re-presentation', *Journal of Management Studies*, **34** (5), 825–41.

Keenoy, T. (1999), 'HRM as hologram: a polemic', *Journal of Management Studies*, **36** (1), 1–23.

Keenoy, T. and P. D. Anthony (1992), 'HRM: metaphor, meaning and morality', in P. Blyton and P. Turnbull (eds), *Reassessing HRM*, London: Sage, pp. 233–55.

Legge, K. (1995), *Human Resource Management: Rhetorics and Realities*, London: Macmillan.

Martin, J. (2001), *Organizational Culture: Mapping the Terrain*, London: Sage.

Mumby, D. and R. Clair (1997), 'Organizational discourse', in T. A. van Dijk (ed.), *Discourse as Structure and Process*, vol. 2, London: Sage, pp. 181–205.

Noon, M. and P. Blyton (1997), *The Realities of Work*, Houndmills: Palgrave Macmillan.

Ohmae, K. (1994), *The Borderless World: Power and Strategy in the Global Marketplace*, London: HarperCollins.

Parker, M. and D. Jary (1995), 'The McUniversity: organisation, management and academic subjectivity', *Organization*, **2** (2), 319–38.

Power, M. (1997), *The Audit Society*, Oxford: Oxford University Press.

Prichard, C. and H. Willmott (1997), 'Just how managed is the McUniversity?' *Organisation Studies*, **18** (2), 287–316.

Reed, M. (2002), 'New managerialism, professional power and organisational governance in UK universities: a review and assessment', in A. Amaral, G. A. Jones and B. Karseth (eds), *Governing Higher Education: National Perspectives on Institutional Governance*, Amsterdam: Kluwer Academic Publishers, pp.181–203.

Shaw, Martin (2000), *Theory of the Global State: Globality as Unfinished Revolution*, Cambridge: Cambridge University Press.

Silverman, D. (1999), *Doing Qualitative Research*, London: Sage.

Smircich, L. (1983), 'Concepts of culture and organizational analysis', *Administrative Science Quarterly*, **28** (3), 339–58.

Taylor, P. and P. Bain (1999), '"An assembly line in the head"': work and employee relations in the call center', *Industrial Relations Journal*, **30** (2), 101–17.

Trowler, P. (2001), 'Captured by the discourse? The socially constitutive power of new higher education discourse in the UK', *Organization*, **8** (2), 183–202.

Watson, T. J. (1994), *In Search of Management: Culture, Chaos and Control in Managerial Work*, London: Routledge.

Wenger, E. (1998), *Communities of Practice: Learning, Meaning and Identity*, Cambridge: Cambridge University Press.

Willmott, H. (1995), 'Managing the academics: commodification and control in the development of university education in the UK', *Human Relations*, **48** (9), 993–1027.

Yin, R. (1989), *Case Study Research: Design and Methods*, Sage: London.

8. Identity Creation in Temporary and Scattered Work Practices in a Relational Perspective

David Metz and Ann Westenholz

INTRODUCTION

The purpose of this chapter is to invite a discussion of how to study the social construction of identities when people participate in 'temporary and scattered' work practices. We argue that the evolution of employee identity and participation in the Danish working life is a complex process from which three different institutionalized modes of ordering have emerged at different times in the twentieth century, that is, orderings of collective bargaining, co-determination and co-management. We assume that the dominant modes of ordering in working life are changing, and that the identities of the institutions – especially companies and unions – are in a position of uncertainty. In this chapter we are interested in understanding theoretically, how this ordering is produced in new configurations of temporary and scattered work practices. In what ways are representations constructed in terms of identities and situations that enable participants to coordinate activities in ways that (temporarily) make sense to them as collectives or individuals? Theoretically we argue for a relational position and we will be drawing on theories, such as communities of practice, knotworking and actor network.

INSTITUTIONALIZED MODES OF ORDERING AND NEW WAYS OF ORGANIZING

In the first part of the twentieth century a collective *system of bargaining* developed in Denmark simultaneously with workers assuming the identity of *wage earners* and *members of workers' collectives* (Lysgaard 1967) *and unions*. Regulatively the development was institutionalized (Scott 1995, pp. 35–7) by the September Agreement of 1899. This agreement confirmed the formal right of the workers to found unions and of the employers to manage work within a certain part of the labour market. In the following years, an increasingly larger part of the workforce (private and public) incorporated and developed this ordering not only as a regulative institution, but also as a cognitive institution (Scott 1995, pp. 35–45). The ordering of collective bargaining was not only institutionalized due to rules governing the labour market, but also became the natural order governing the relationship (cognitively) between the two sides of industry. Being a cognitive institution implies that the parties found it natural to view each other as performing the roles of wage earner and manager, respectively. This became part of their identities. Since World War II and especially since the 1980s the tendency has been to ascribe increasing importance to decentral negotiations in the individual company, and to include more issues. Despite these changes, the ordering is still based on the idea that workers participate in working life as wage earners in relation to the company, and as members in relation to a union. And the fundamental principles in terms of the employer's right to manage work have not been changed significantly (Dansk Arbejdsgiverforening 2000, p. 178; Due, Madsen and Strøby Jensen 1994; Knudsen 1995; Scheuer 1996; Westenholz 1994). In the vein of Friedland and Alford (1991, p. 248) the collective ordering of bargaining can be said to draw on an institutional logic that can be characterized as the logic of capitalism. The institutional logic is a set of material practices and symbolic constructions that constitute the organizing and is available to organizations and individuals for further elaboration. The capitalist way of organizing is accumulation and commodification of human activities.

Since World War II a different mode of ordering has emerged – *co-determination* – in which employees were/are constituted through membership of *two* organizational units. First, employee membership of unions are maintained (in the 1950s approximately 60 per cent of the labour force was unionized; today approximately 88 per cent is unionized (Dansk Arbejdsgiverforening, 2000)). Second, employees are allocated a new membership of the company in which they work. Employees were/are guaranteed the right to discuss issues of common importance with management through shop stewards, and companies develop closer ties with employees. This dual membership can be said primarily to draw on an institutional logic of capitalism in that the original agreement between unions and employers' associations is still valid. Today, however, an element of employee participation has been incorporated into the decision system of the company that is rooted in a different institutional logic – that of democracy. The mode of ordering of democracy is participation and common control of human activities (Friedland and Alford 1991, p. 248; Westenholz 1994).

Since the 1970s a third type of ordering has emerged – *co-management* – in which solely the company, and not the unions, has been socially constructed as the place where employees demonstrate loyalty and where their collective identities are created. Inspired by Selznick (1969), who developed the concept of 'the organizational citizen', Christensen and Westenholz (1999) and Westenholz and Christensen (2001) call this identity creation *organizational citizen*. An example of this development is the law on employee representative on the boards of directors in Danish limited companies that Christensen and Westenholz analyse by drawing on an institutional logic of democracy (Friedland and Alford 1991, p. 248).

Collective bargaining, co-determination, and co-management are institutionalized modes of ordering that coexist as both regulative and cognitive institutions. In this context we will not discuss in which ways or how the modes of ordering are distributed in working life, but merely mention that sometimes they are all carried by single individuals and sometimes they are all three identifiable in the single company (Westenholz 1999).

We assume, however, that the three modes of ordering are subject to pressure, and that we are witnessing destabilizing of institutions in that we can observe new ways of organizing working life. In these new ways of organizing it no longer makes sense for the participants in their daily practice to draw on the symbolic meanings of unions, companies, or their own role as wage earners or citizens of the company. Employees are increasingly participating in practical work relations that break with established perceptions of the significance of unions and companies, and thus also with the importance of wage earner identity and company identity. Viewed from the perspective of unions and companies a new type of employees is emerging that is defined as participants in relations of work designated 'temporary and scattered' compared to what we are used to not only in Denmark but also in other European countries and in the USA. (Barley 1996; Barley and Kunda 1998; Engeström, Engeström and Vähääho 1999; Garsten 1999; Nardi, Whittaker and Schwartz 2000; Østerlund 1996) Such employees are sometimes referred to as 'atypical' (Garsten 1999, p. 606; Teknologisk Institut 2000).

We are not arguing that such work relations are unprecedented, or that they are/will be dominating working life in the next couple of years. But we assume they will grow in number and that they are possibly pointing toward a trend in working life that will be more significant in the future. Let us illustrate this with a few examples.

Within a number of companies (for instance, IT, consultancy, biotechnology, and entrepreneur firms) we see employees (product developers, sales people, service providers) that are employed in one company, but increasingly collaborate with employees in other companies, with freelancers or with researchers at universities. Sometimes they are placed geographically in a company other than the one in which they are employed. They are not only crossing the physical boundaries of the companies, but probably also the mental ones. Our preliminary term for this kind of employees is the '*boundary spanners of the company*'. Also, in a number of different company types the tendency is for some employees to have workstations at home, meaning that in a smaller or larger share of the working hours their relationships with the workplace are *virtual*. We are witnessing a growth in freelancers not only among traditional lines of business such as journalism, but also in information and communication technology, which we find especially interesting in that they are a new and rapidly growing group both as a trade and as individuals. First, this means that they have no historical antecedents to draw upon in negotiating identity(ies) as opposed to the case in other fields characterized by this type of work, for instance, journalists and more traditional crafts such as masonry. Second, the technology involved plays a unique role. It is inherent in the concept of 'computer nerd' that individual social competencies are coupled with personal computers (PCs) in ways that differ from other fields of work, and the pace of the technological development implies that education and qualifications soon become obsolete. The final example is what we call *migrants in the labour market*, that is, people working on short-term contracts in companies. They may be engineers or other types of technicians that are hired to solve specific problems, and having done so they move on to the next job in another company.

We assume that employees who cross institutionalized boundaries through participation in temporary and scattered work practices find themselves in ambiguous situations as to their own identity and the identity of those with whom they collaborate. Only a few empirical studies have focused on this phenomenon, and the purpose of this chapter is to invite a discussion of how we theoretically can understand the development of identities in these special type of work practices. We will illustrate our points with examples from our empirical research.

A Relational Meta-Scientific Position

The research question that we are resolved to understand starts with the questions of how we as researchers establish a population of people that are participating in temporary and scattered work practices. Such people will often be characterized as 'atypical' or 'exposed'. The choice of terms points towards that we (labour market researchers, unions and sometimes also companies and the involved employees) find these people to be in an uncertain and ambiguous situation as far as identity (and other issues) is concerned. They are viewed as 'without stable position or affiliation with a company or a union' in their career. Uncertainty and ambiguity are assumed to emerge because these people in their work practices are moving across or away from institutionalized modes of ordering – in this case 'the company' and/or 'the union'. However, we will argue that it is just as legitimate to claim that the dominant modes of ordering in working life are becoming deinstitutionalized, meaning that it is the identity of these institutions that are being placed in an uncertain and ambiguous position. We wish to understand how order is being (temporarily) produced in these new work practices, and how it can be represented in identities and situations thus enabling participants to coordinate activities in ways that (temporarily) make sense to them as collectivities or individuals.

Below we will position ourselves meta-theoretically in relation to the question of how we should understand identity. In the following section we will describe how we will approach this question analytically in our empirical investigation.

The foundation of our analysis is a relational paradigm. This position presumes individuals to assume character through the relations that they are part of. The individual cannot define himself or herself without drawing on a socially constructed understanding of the type of situation, the person is in or which role that s/he plays in relation to others in the 'game'. People are not billiard balls, but rather 'play dough' balls that take shape/are shaped in the relation. At the same time, the relation will not shape the person in a deterministic way. If this were the case, the situation would be one of norm-following, and the participants would a priori agree on the nature of the situation and which role they should play (a priori common logic of appropriateness). Against this we assume that participants are – and most often will be – uncertain or ambiguous as to establishing the type of situation or their own role, and this uncertainty or ambiguity is settled through negotiations[1] among the participants. We agree with Emirbayer (1997, p. 289) that 'what is distinct about the transactional (relational) approach is that it sees relations between terms or units as preeminently dynamic in nature, as unfolding, ongoing processes rather than as static ties among inert substances', and he continues citing Dewey and Bentley (1949, p. 133) 'no one would be able successfully to speak of the hunt*er* and the hunt*ed* as isolated with respect to hunt*ing*. Yet it is just as absurd to set up hunt*ing* as an event in isolation from the spatio-temporal connection of all the components'. This quotation reflects our very point: identities become established when subjects are recognized in relations to others, but recognition is 'negotiated' among the subjects in a *dynamic unfolding process*. Wenger (1998, p. 151) puts it this way:

> Identity in practice is defined socially not merely because it is reified in a social discourse as a lived experience of participation in specific communities. What narratives, categories, roles, and positions come to mean as an experience of participation is something that must be worked out in practice. ... In the same way that meaning exists in its negotiation, identity exists – not as an object in and of itself – but in the constant work of negotiating the self.

Consequently, identities cannot be viewed as exogenous variables, but must be endogenous in the analysis. The concept of identity cannot be used to explain what happens, but it can be used to describe what *has* happened.

Unfolding the Relational Paradigm through Three New Theories

In the remains of this chapter we shall draw upon theoretically related novel traditions that have dealt with similar empirical problems. Here we will point to three traditions that each in their way contributes to elucidate the research question. We are only able to give a very brief presentation of the theories and short sketches of analyses of empirical case vignettes from our ongoing research. The cases presented in the following are prepared after a two-month field study stay with participant observation at a web company supplemented with a number of individual interviews. We do not claim representativity of these data. They are first and foremost illustrative of our main argument that established institutions are under pressure, and they can guide the focus for further studies by giving examples of in which way that pressure is applied through changes in practice.

COMMUNITIES OF PRACTICE (SITUATED LEARNING)

This approach really started with Jean Lave and Etienne Wenger's (1991) book *Situated Learning*. Here, the two authors develop a theoretical perspective on learning based on the idea that learning must be understood as *participation in communities of practice*. Since then the concept of communities of practice has, among other things, been criticized for being imprecise or downright unclear. Jean Lave is now only applying the term descriptively (cf. Lave 1999) whereas Etienne Wenger has attempted to develop it into a more applicable analytical category (Wenger 1998).

In our reading of Wenger, 'communities of practice' becomes a basic concept for a series of features characterizing the social contexts in which human beings participate, including their working lives. A community of practice is thus not identical with a workplace or an organization. It is, rather, a new way of perceiving working life that attempts to break with traditional ideas about fixed structures governing or constituting the framework for everyday practice. Our interest centres on this new take on drawing boundaries and the ensuing consequences for the understanding of identity.

Wenger applies two concepts to capture the relation between process and structure: participation and reification. Participation is the continuously ongoing process of acting in a social context. Reification indicates that participants occasionally view the ongoing processes as objects and hence as more stable in time and space. On the one hand, there is the participation aspect of working life, which is what we actually do, the concrete actions that we perform. On the other hand, there is the reification aspect, which is what we do when observing our (joint) participation as objects. Reification becomes manifest, for instance, in work being viewed as a specific 'job' identifiable in terms of a 'job description'. Thus boundaries are socially constructed as a consequence of people doing things together in mutually recognizable enterprises. Identities are likewise consequences of ways of engaging in practices, thus in Wenger's terms conceptualized as 'a nexus of perspectives across boundaries'.

Let us illustrate the interplay between participation and reification, boundaries and identities in communities of practice by an example from our ongoing research. A small company that develops web applications decides to develop a 'help-desk' application aimed at on-line banking. They split up the task and Victor is assigned a specific subtask for which he has special qualifications – the development of a chat function as an integrated part of the help-desk. Victor has previously done all the work in the company setting up both the internal network and the Internet servers, so it seems natural to everyone that he should do the part that requires specialized knowledge of network programming. In fact some of the others even refer to him as a super network-geek.

Since Victor's programming has to be integrated with that of some of the others, they discuss which programming technology Victor should use for the chat function. Victor argues that they should use *open sockets* since this is a way of doing it that results in a fast working solution that will require a very low amount of resources from the total system, thus making it a technologically 'good' or even 'beautiful' way of doing it. The other programmers know about programming open sockets, but they are not familiar with using it 'hands on'. They originally had another solution in mind but are convinced by Victor's arguments that the open sockets solution is superior. After a while Victor is ready for a presentation of the chat function. He installs the application on the company's internal network and the chat function works fine. The next step is 'bug-hunting', that is, searching for errors in the programming of the whole application. The bug-hunting is done by some of the other programmers and quite a few errors show up as they go along. One thing surprises them, though; they discover that the chat function does not work at all when there is an active firewall.[2] The bug hunters talk to Victor about it and he says that open sockets only works if the firewall is down. The others argue that it is not an option to require a bank to shut down their firewall, but Victor still maintains that he has implemented the best solution.

This description features a couple of examples of reified identities. Victor is *a programmer* in the company and *a super network-geek*. These two 'titles' are signifiers of reification. Behind the signs lies an understanding of what it means to hold these titles, not necessarily in society as a whole, but at least in the local community of practice. The titles are locally meaningful examples of talking about *who* or *what* someone is in reified terms. When we look at what kind of participation holding the titles entail, we see that being a programmer among other things is about sitting in front of a computer writing many lines of code, sometimes testing parts of it, sometimes asking colleagues for advice on how to solve logical problems or what the correct annotation for a function call is, most of the time working alone. Being a super network-geek is a relative term implying a special status that has to do with a niche in programming and computers – a niche, that is, in this context. In practice it is, among other things, about drawing cables, working at many different computers, talking on the phone to outside service providers.

Victor the programmer seems to fit in comfortably in the environment. He is recognized as a legitimate participant qua programmer or doing programming. It is important to emphasize here that it is not reducible to a question of position, function or formal title which might be implied by this way of putting it. The legitimacy emerges out of the daily practice in the company, which is recognizable as a joint enterprise of the participants sharing a repertoire of how to do their practice in mutual engagement.

The three terms 'joint enterprise', 'shared repertoire' and 'mutual engagement' are dimensions or analytical categories in Wenger's understanding of a community of practice. This is what plays a part in boundary construction. So, as a programmer, Victor is engaging in an enterprise with other programmers using some of the same tools and language.

But Victor is also using a technology he does not share with the others. In the example we emphasized *open sockets* which reifies a view of the activity of making an open communication-line between two or more computers in a certain way. That is, *open sockets* is a specific set of tools that one must be a knowledgeable user of in order to be acknowledged as a legitimate member of a community. As is apparent in the example, Victor is the only one in the company with knowledge of open sockets, so in a way he thus represents a relative outsider in the local context. He is still a legitimate participant through his earlier contribution of setting up the company network, but when it comes to cooperation with the others something fails. One way of putting it could be to say that Victor is too much of a network-geek and too little of a web programmer in his way of working with the chat function. As one of the founders of the web company put it later on: 'After that experience we went for hiring people with a more pragmatic attitude.'

In terms of identity-as-participation this aspect of the example shows that Victor is also part of another community, one we might label 'network specialists'. The different memberships and corresponding identities can be illustrated as in Figure 8.1. This is very much a simplification of Wenger's concepts for illustrative purposes. The idea is to show that communities of practice (CoP) are not simply yet another new way of cutting up an organization structure, since the communities extend beyond the boundaries of the company as shown in the combined final illustration.

The first illustration shows Victor's participation in the 'programming-CoP' making web applications. In practice this CoP extends beyond the borders of the organization. For example, some of the programmers use open source software and thus refer to a global network of programmers outside the company. The second illustration shows Victor when he sets up the network of the company, as such participating in a broader CoP of IT-users in the web company. The third illustration shows what happens in the second part of the description when it becomes apparent that Victor besides being a programmer is as well a network geek. He participates in a CoP of network specialists which no one else in the company does. Analytically we can say that, although his job seems to be one of brokering between the two CoPs, it seems as if he ends up being more of a network-configurer (geek) than a programmer. From the point of view of the programming-CoP the network-configuring is not successfully integrated in his participation. From Victor's point of view, as illustrated last, however, it seems fair to say that his programming is successfully

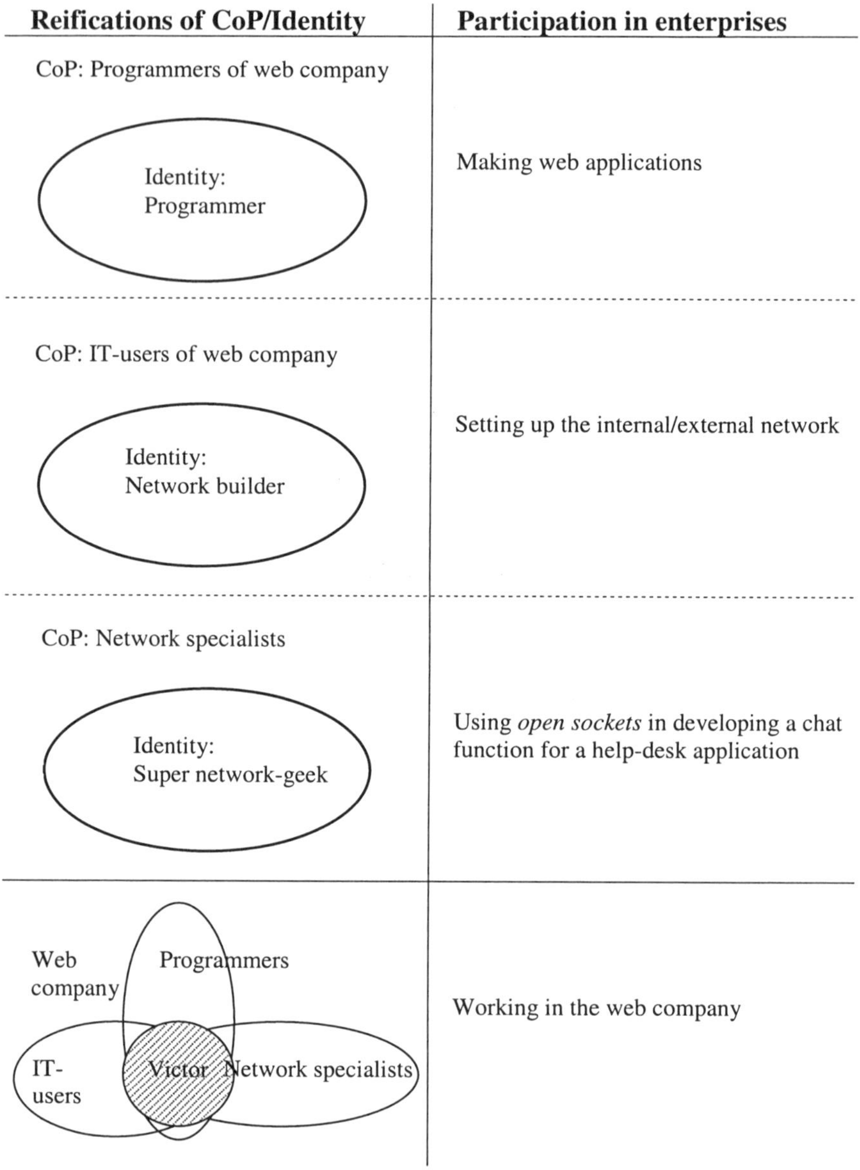

Figure 8.1 Identities as participation in communities of practice

integrated in the network-configuring CoP – regrettably this is not useful to the company, though.

In other words, what happens during Victor's stay in the company is that he participates in different communities of practice brokering between them, especially with respect to the use of specific technologies. To some extent the brokering is successful, giving rise to the legitimate identities of 'programmer' and 'network builder'. In the end the brokering fails or becomes unidirectional[3] and gives rise to a non-legitimate identity of 'geek'. Legitimacy is here understood as a question of successful negotiation of crossing the specific boundaries between the involved communities of practice.

We have now, to a certain extent, specified the empirical focus, which is relevant for investigating identity in working life from a relational position. This perspective has shown us that the types of boundaries implied in the three modes of ordering we identified in the beginning of the chapter are not sufficient, or perhaps they are even misleading, in studying this type of practice. But we need to take a step further than Wenger who places negotiations of identity in communities of practice. In relation to our project the problem is that loose couplings characterize the working lives of the participants, whom we wish to investigate. It seems tempting to merely assert that this 'type' of participant encounters problems with their identities and practices as might easily be implied in this example, clearly indicated in the term 'geek', but in our view this would be a mistake since that would practically render everyone we wish to look at a problem. We feel that Wenger's concepts are more easily applied to work practices where it seems more or less unproblematic to talk of established boundaries. His own examples are from a large insurance company where established boundaries – internally between departments and externally towards customers – play an important part. On the other hand, we are certain that the concepts were not intended to be reserved for that type of practice, but they tend to carry images of it with them. Thus we have chosen to look for further perspectives in developing the theoretical position. First we introduce the theory of knotworking.

Knotworking

The concept of knotworking derives from activity theory, which is a continuation of the cultural-historical school in Russian psychology (cf. Engeström 1987). For Engeström and colleagues (1999) knotworking designates a type of practice ('activity') that is not organized around a centre. 'Knotworking' must be understood as a metaphor for a pulsation of activity without a predetermined goal or object, in which agents participate at irregular intervals and without performing certain fixed roles. It is a kind of loosely coupled ad hoc organizing.

Consequently, a knotworking analysis deals with empirical phenomena similar to those we are focusing on. Although Wenger at some length discusses questions of delegation, mediation and brokering between, across and on the boundaries of communities of practice, we find it more helpful to use a perspective that primarily focuses on these phenomena.

Let us take a look at the web company again. After the general crisis in the dot.com business, they have decided to scale down and instead make more use of freelancers. Some are former employees and some are new. They offer some of the regular freelancers office facilities, which gives the freelancers a base to work from and the company have them close by if they need them for assignments. The latest idea has been to join forces with another company that also uses programmers. They agree that the programmers from the other company move location to the web company's premises so the programmer's from the two companies can more easily make use of each other. On the day the 'new' programmers move in, everyone from the web company is present, as well as two freelancers: one working on a job for the web company and another working on a job for a third party.

The new programmers say hello, plug in their computers and start working. After little more than an hour the power disappears from most of the office space. Everyone looks up more or less puzzled. One from the web company starts asserting the spread of the power-loss and identifies which fuse has blown but cuts the power from all outlets in his search. After some discussion they decide the power-cabling in the office must be redone to distribute the consumption on outlets from different fuse-groups. The manager turns off all electrical equipment in his office to allow the programmers to work while others draw a new cable from a less exerted fuse-group. Since all power has been cut off, the central servers and the router to the Internet connection has to be restarted to get the network back up. The new programmers also need the Internet connection in order to work on the servers in their own company. One of the freelancers, a new one called Derek, immediately goes to work on the servers. After putting them back up he returns to his seat to configure them from there. He talks to a regular employee to obtain usernames and passwords and to the other freelancer present, who happens to be a former employee, to discuss some of the technicalities of server configuration.

Applying a knotworking-inspired analysis will enable us to spot the new aspects of identity negotiation that emerge in the literally ad hoc organized work described in this example. Engeström and colleagues call this co-configuration, that is, a mutual shaping of relations among participants, the object of practice, and the larger context that they are part of.

Disturbances occur in the system of relations during the described process, disturbances that change the object of practice and the participants. At the outset of the vignette everybody goes about business as usual. A couple of freelancers are present, but it is clear to them and everyone else what they do and why. Then the relation between participants and tools are disturbed, the computers are brutally shut down. A need for reorientation of the object of practice arises which affects each participant's position both towards the others and towards the tools at hand. What we see in response is a fluctuation of identities among participants.[4]

This ideally requires a longer analysis than we wish to present in this context, but if we limit our focus to Derek, he changes from freelance-programmer to network-maintainer even though he is new and not even a regular employee. As the main work tools suddenly cease to work the whole system of interrelationships between participants is disturbed. So what happens in response cannot be understood as simply individual or even collective sense-making. What happens is not just a reorientation of how to deal with the situation, but a reorientation of the relative positions of participants. Derek must revise his object of practice and thereby go beyond the agreed division of labour – the job he has been hired for. In fact everybody seems to do this kind of revising, but the interesting thing is that while the regular employees' reorientation could be explained in terms of their membership of a common community, as in the previous case, this cannot explain what Derek does. Derek is an outsider to the community and yet he takes part in resolving the situation as if he was engaged in the same enterprise as the others. Why this happens cannot be accounted for within this framework, but we may try to understand how it comes to happen. One important factor is the way work in general is organized in the company, which can be described as a mixed division of labour where people are given individual responsibilities based on their recognized qualifications or areas of expertise combined with a general expectation that whoever identifies a problem tries to solve it. Derek expresses this while talking to another freelancer during a social event:[5]

> I used to work for a major Internet access provider where I worked my way up from customer support back when they were still just a small company consisting of enthusiastic young people. When they were bought up by a large foreign company things changed. For instance I once found a spelling error on the company homepage but to fix it I had to contact the webmaster higher up in the

> ranks – I did not have access to edit the homepage. On the other hand the same thing happened after I had been here for a week, and they just gave me the administrator password so I could fix it myself.

So in a way it can be said that this shifting of identities is not particularly out of the ordinary in this context, it is just obviously apparent to an outsider due to the dramatic break down of ongoing practice.

One of the primary gains from a knotworking-inspired analysis for our purpose is that in order to understand how this practice develops and takes shape it does not suffice to search for specific foundational premises that practice either rests on or stems from. Instead, we must attempt to look at the relations that become established and how these relations continuously determine what they are relations between. As Engeström and colleagues put it: 'the center does not hold', it is not possible to define a single regulatory principle or entity in the constitution of practice or identities. Rather the opposite seems appropriate to say: many different entities and principles interrelate and out of this 'mess' emerges an image of pulsating identities. So again we can ascertain that the initially identified modes of ordering are insufficient to establish an understanding or explanation of what is going on in this way of working. Furthermore we claim that a communities of practice perspective tends to have a bias towards more stable types of relations where the knotworking seems better suited to grasp more transient and dynamic ways of relating.

But there are reasons for continuing our search for appropriate theoretical frameworks that can guide and capture our empirical attention. In both the recounted examples technology plays an important part in what goes on, but it is only given a passive or indirect role in the analysis of identities. It seems puzzling that what is apparently prominent in the descriptions becomes secondary in analysis. This asymmetry has led us to take up actor-network theory.

Actor-Network Theory

The theory of actor-networks (ANTs) has developed from studies in science and technology and is best described as a series of methodological principles that attempt to fit into the generalized principle of symmetry (Callon 1986; Latour 1993; Law 1994). In the original formulation of the principle of symmetry (cf. Bloor 1976) the point is that scientific successes and failures must be explained using identical principles. For instance, a successful scientific experiment cannot be explained by referring to its reflection of the order of nature, and a failed experiment by referring to the sloppiness of the laboratory technician. These are asymmetrical explanations that implicitly alternate between privileging Nature and The Social as causes of success and failure. Actor-networks theory takes this point to its extreme consequence and asserts that nothing can be privileged a priori, but everything must be explained from identical premises. An aeroplane crash, for instance, cannot be explained by human error simultaneously with asserting that aircrafts do not crash because of efficient technology. The fact that aircrafts both crash and do not crash must be explained in 'the same way', that is typically as a mixture of, for example, technology and humans. This is what ANT does by operating with a somewhat alien terminology inspired by semiotics. Below we will present a few examples.

The term *hybrid* designates the fact that entities in a given context are best described as composed of different types of materials also called *heterogeneity*. When action potency is attributed to entities in a description of what is happening in a situation, they are also termed *actants*, that is, 'that which does something'. In most cases actants are best perceived as hybrids that emerge as an effect in heterogeneous *networks*. That is, when we single out some people as being the actors in a given context, we must see them as having emerged from the relations among the dynamics of diverse materials. In consequence, the ANT views subjectivity, that is, to have status as an acting agent, as being constituted through concrete relations among diverse materials in practice. This is as opposed to subjectivity, being something that a priori can be assumed to belong to specific entities, typically individuals. The latter is the assumption within the approaches of both communities of practice and of knotworking.

In relation to our argument, the ANT approach will draw attention to two things. First, how subjectivities are constructed in the links between technologies and other materials, and second, that identities are versatile – actors have more than one identity, they have several 'at once' that interact in a network of relations. We can exemplify these ANT attentions by briefly returning to the two former examples.

In the community of practice example Victor is a knowledgeable user of the open sockets technology. Applying the ANT sensibility we can analyse events differently by giving the technology agency in our analysis. In this perspective the open sockets have obviously enrolled Victor as a spokesperson, or rather the hybrid 'Victor-open sockets' seems to be a better description of who or what is at play in the vignette. This actant is working towards a specific ordering commensurable with the larger heterogeneous network which it is a part of – a mode of ordering we could call 'techno-logic'. The thing is that this 'techno-logic' dictates among other things what constitutes a 'good or even beautiful solution'. As events unfold this, however, proves not to be a 'pragmatic' solution, that is, it does not conform with another actor-network with that mode of ordering. The other network consists of hybrids of programmers, banks, firewalls and the web company's 'bible': *The Pragmatic Programmer* (Hunt and Thomas, 2000). In the description the firewall and the bug tester plays the important part as the hybrid spokesperson of the 'pragmatic' network – doing the exclusion of 'Victor-open sockets' from the pragmatic network.

What we gain from this analysis is a first hint at *why* events turn out as they do. To answer the 'why' question the analysis would have to be developed further than we do here, though, but the answer lies in the direction of network sizes, distances and the local number and power of hybrid actants. Further, we now see how non-human or hybrid actants participate in the construction of what goes on and that the outcome identity 'geek' attributed to Victor is not just Victor's but one given to the hybrid of which Victor is 'a part'. Victor, however, also 'shares' other identities with the open sockets technology – as a spokesperson for the techno-logic network he is an 'expert', and as the insistent arguer of choosing this technology he is a purist or an idealist. If we look at discourses on for instance knowledge management or supply chain management it is apparent that some of these other identities could have emerged as more prominent in the story, if only the techno-logic network had been able to mobilize more or stronger actants in the context.

In the knotworking example a similar analysis can be made. A technical actor-network consisting of power supply, client and server computers, Internet router, cabling and outlets is very prominent in shaping events. Derek is enrolled or lets himself be enrolled in it in a new way, but this is simply a part of an ordinary operation of the actor-network. It is a fluid network as is apparent in the statement Derek makes about gaining access to the company web site. This fluidity has a dual effect, though: on the one hand, it adds plasticity to the overall network rendering it less susceptible to the potentially detrimental consequences of 'accidents'; on the other hand, plasticity becomes a functional requirement disciplining enrolled actors. Perhaps Derek is more prone to adhere to this requirement because he is 'new'. This is not a point about mere social pressure; it is not that it seems to be especially explicitly or implicitly expected by the other individuals present that Derek uses this opportunity to 'prove his worth in the company'. The point is that Derek has to do the job of getting the system back up, just as everybody else does, only he in effect happens to be the one that responds the quickest, perhaps because of a heightened sensibility to this demand coming from being 'new'. But it probably also has to do with him being a freelancer as opposed to Victor in the other example who proves to be everything but 'fluid'. Derek is not strongly connected to another network, he is not a spokesperson. So Derek is not an 'expert', he is an 'opportunist' in the neutral sense of the word – one who catches the opportunity as it arises – just as everybody else in the company apparently does.

We will need further empirical data to analyse how the unfolding of these practices can be characterized in terms of modes of ordering, but it seems clear that we have evidence for a breach in the historically established ones.

CONCLUSION

In the chapter we have argued that new forms of temporary and scattered practices in working life are evolving. This is taking place within many areas, such as, for instance, information technology and biotechnology. Employees are increasingly crossing company boundaries in their cooperation, by working as freelancers or migrants in the labour market. Such employees are often referred to as 'atypical' or 'exposed' because they are perceived as being without stable positions compared with traditional institutions such as unions or companies. However, we would rather argue that the dominant ways of organizing working life are currently being de-institutionalized. This implies that for some employees identity in working life is not necessarily rooted in membership of a union and hence as 'wage earners' – or as employed in a company with the identity of 'citizen of the company'. Identity and affiliation is not something one is/has, but something that one develops in being recognized in *relation* to others – in relations where recognition is 'negotiated' among the involved actors in temporary and scattered work practices. Such 'negotiations' are characterized by temporary reductions in the ambiguity of who the involved 'are'. What we have argued in this chapter is that three supplementary perspectives are useful in analysing these negotiations: communities of practice supplies an alternative view on *boundaries* than that of, for instance, companies and unions. Knotworking helps us understand relations characterized as temporary and scattered, while ANT brings in an understanding of the role of technology in the process.

We can summarize our understanding under four points. First, we propose to see identities as endogenous variables in the analyses and not as exogenous psychological or culture sociological variables. This means that identities are not taken as given in analyses, but must be seen as developing. Second, we do not perceive identity development as something tied to the individual – psychologically or culture sociologically – but as a relational phenomenon or in the words of Mead (1934) 'selves can only exist in relationship to other selves'. Third, we suggest that identity is not necessarily rooted in the relation among predetermined entities (typically individuals), but quite the contrary identity is created in the relation between technologies, humans and diverse other materials. Fourth, identities must be understood in plural, meaning that an actor has more than just one identity – s/he has several identities that interact in a network of relations. By this we hope to be able to contribute to demystify or normalize a group of employees that are currently referred to as 'atypical', 'expelled', 'threatened' or the like and open up for an empirical investigation of organizing of identities in temporary and scattered work practices.

NOTES

1. We use the term 'negotiation' here as a term covering the social process where people acting together establish relational identities. This is as opposed to the sense where actors are bargaining their respective interests, as this latter meaning presupposes identities established beforehand.
2. A 'firewall' is a piece of software or hardware that prevents external users from accessing an internal network – a protection against malicious crackers (computer-criminals) used by all corporations and lately an increasing number of private users. It is a sort of damage-control technology much like a 'real' firewall.
3. Unidirectional since the web application is subsumed into the enterprise of the CoP of network specialists and not vice versa.
4. Engeström does not employ a concept of 'identity'. We have read his analysis of community embedded subject positions as analogous to our use of identity.
5. This is not a direct quote but a comprised version of what he said.

REFERENCES

Barley, S. R. (1996), 'Technicians in the workplace: ethnographic evidence for bringing work into organization studies', *Administrative Science Quarterly*, **41**, 404–41.

Barley, S. R. and G. Kunda (1998), 'Bringing work back in', paper presented at the SCANCOR Conference 'Future of Organizations', Stanford, USA, September.

Bloor, D. (1976), *Knowledge and Social Imagery*, London: Routledge and Kegan Paul.

Callon, M. (1986), 'Some elements of a sociology of translation: domestication of the scallops and the fisherman of Saint Brieuc Day', in J. Law (ed.), *Power, Action and Belief: A New Sociology of Knowledge? Sociological Review Monograph*, London: Routledge and Kegan Paul, 32, pp. 196–233.

Christensen, S. and A. Westenholz (1999), *Medarbejdervalgte i danske virksomheder – fra lønarbejder til borger i virksomhedssamfundet*, København: Handelshøjskolen Forlag.

Dansk Arbejdsgiverforening (2000), *Arbejdsmarkedsrapport 2000*, Dansk Arbejdsgiverforening.

Dewey, J. and A. F. Benley (1949), *Knowing and the Known*, Boston: Beacon Press.

Due, J., J. S. Madsen and C. Strøby Jensen (1994), *The Survival of the Danish Model: A Historical Sociological Analysis of the Danish System of Collective Bargaining*, København: Jurist- og Økonomforbundets Forlag.

Emirbayer, M. (1997), 'Manifesto for a relational sociology', *American Journal of Sociology*, **103**, (2), 281–317.

Engeström, Y. (1987), *Learning by Expanding – An Activity-theoretical Approach to Developmental Research*, Helsinki: Orienta-Kosultit Oy.

Engeström, Y., R. Engeström and T. Vähäaho (1999), 'When the center doesn't hold: the importance of knotworking', in S. Chaiklin, M. Hedegaard and U. Juul Jensen (eds), *Activity Theory and Social Practice: Cultural-Historical Approaches*, Aarhus: Aarhus University Press, pp. 344–75.

Friedland, R. and R. R. Alford (1991), 'Bringing society back in: symbols, practices, and institutional contradictions', in W. W. Powell and P. J. DiMaggio (eds), *The New Institutionalism in Organizational Analysis*, Chicago: University of Chicago Press, pp. 232–63.

Garsten, C. (1999), 'Betwixt and between: temporary employees as liminal subjects in flexible organizations', *Organizational Studies*, **20** (4), 601–17.

Hunt, A. and D. Thomas (2000), *The Pragmatic Programmer*, Boston: Addison-Wesley.

Knudsen, H. (1995), *Employee Participation in Europe*, London: Sage.

Latour, B. (1993), *We Have Never Been Modern*, London: Harvester Wheatsheaf.

Lave, J. and E. Wenger (1991), *Situated Learning: Legitimate Peripheral Participation*, Cambridge: Cambridge University Press.

Lave, J. (1999), 'Læring, mesterlære, social praksis', in K. Nielsen and S. Kvale (eds), *Mesterlære*, København: Hans Reitzel, pp. 33–5.

Law, J. (1994), *Organizing Modernity*, Oxford: Blackwell.

Lysgaard, S. (1967), *Arbeiderkollektivet*, Oslo: Universitetsforlaget.

Mead, G. M. (1934), *Mind, Self and Society from the Standpoint of a Social Behaviorist*, Chicago: University of Chicago Press.

Nardi, B. A., S. Whittaker and H. Schwarz (2000), 'It's not what you know, it's who you know: work in the Information Age', *First Monday*, **5** (5), May (http://firstmonday.org/issues/issue5_5/nardi/).

Østerlund, C. (1999), 'Salgslærlinge på tværs', in K. Nielsen and S. Kvale (eds), *Mesterlære – læring som social praksis.* København: Hans Reitzel, pp. 147–61.

Scheuer, S. (1996), *Fælles aftale eller egen konkrakt i arbejdslivet*, København: Nyt fra Samfundsvidenskaberne.

Scott, W. R. (1995), *Institutions and Organizations*, Thousand Oaks, CA: Sage.

Selznick, P. (1969), *Law, Societies, and Industrial Justice*, New York: Russell Sage.

Teknologisk Institut, Erhvervsanalyser (2000), *AC'ere i atypiske beskæftigelsesrelationer – Sammenfatning*, København: Teknologisk Institut.

Wenger, E. (1998), *Communities of Practice – Learning, Meaning, and Identity*, Cambridge: Cambridge University Press.

Westenholz, A. (1994), *Modeller af arbejdspladsdemokrati efter 2. verdenskrig. Fra et logik- til et paradoksperspektiv med fokus på det organisatoriske niveau*, Doktorafhandling, Handelshøjskolen i København.

Westenholz, A. (1999), 'Medarbejderdeltagelse i ledelsen af arbejdslivet – en historisk analyse med fokus på danske forhold', in S. Christensen and A. Westenholz, *Medarbejdervalgte i danske virksomheder – fra lønarbejder til borger i virksomhedssamfundet*, København: Handelshøjskolen Forlag, kap. 2: 23–38.

Westenholz, A. and S. Christensen (2001), 'Employee representatives as company strategic actors – in an enacted world', *Arctic Research Journal*, (1), 146–67.

9. Identity Construction and Commitments of Graphic Designers

Ellen van Wijk

INTRODUCTION

Advertising agencies in the Netherlands are forced to become more innovative and flexible due to the rise of new technologies such as the Internet, internationalization of the client market, and a growing professionalism of the clients (Susijn 1999). One way in which advertising agencies attempt to cope with these developments is through employing specialized professionals. However, agencies are confronted with high mobility rates and a tight labour market, and these circumstances have raised their interest in new ways to bring about greater commitment, especially organizational commitment, on the part of staff. Apart from the positive impulse organizational commitment is assumed to have on achieving organizational goals, it is assumed that employees who are committed to the organization stay longer in the organization (Mowday, Porter and Steers 1982). The question is whether this wish for organizational commitment of the management fits the meaning which employees attach to commitment.

This chapter focuses on the meaning which graphic designers, as a specific category of creative professionals, attach to commitment and on the origin of these commitments by exploring the construction of the identity of graphic designer. The special interest in graphic designers arises from the fact that two-thirds of the creative professionals employed by advertising agencies, including graphic designers, have considered working as a freelancer (van Wijk 2000). This outcome appears to confirm literature which suggests that professionals are more committed to their profession than to the organization they work for (Gastelaars 1997). However, not all creative professionals actually take the step of becoming a freelancer. This situation raises interesting questions concerning the decision-making process and the commitments of graphic designers and of creative professionals in general. Could it be that creative professionals experience professional and organizational commitment simultaneously and, if so, what is the relationship between organizational and professional commitment? Are other cues of influence in the decision-making process?

These questions will be addressed using empirical data on graphic designers in a design agency which were collected in open interviews.

This chapter is organized along the following lines. First, a theoretical framework is sketched in which the concepts used are elucidated, followed by a short exploration of related research on professionals and their commitments. The subsequent sections describe and analyse empirical data on the identity construction of graphic designers and their commitments. The last section presents some conclusions and points for further discussion.[1]

THEORETICAL PERSPECTIVE

Organizational commitment has been given a great deal of attention in recent organizational theories. Literature on these theories reflects the ongoing debate on the meaning of the concepts of organizational commitment and organizational identification that are both being used (cf. Whetten and Godfrey, 1998). Part of the discussion is whether commitment and identification are different names for the same concept – as suggested by those who favour the attitudinal conceptualization of organizational commitment – or not – as suggested by those who conceptualize organizational commitment in economic terms (Pratt 1998). This controversy is too complex to discuss here, but it shows the need to clarify the use of the concepts of commitment and identification in this chapter.

In the literature, identification is a confusing concept which can refer to an outcome, a feeling of sameness of the individual with the perceived identity of the object (cf. Tajfel 1981), as well to the process of identity construction (cf. Vermeulen 2001).

Identification as the outcome of a process is commonly used. Tajfel (1981) defines identification as part of an individual's self-concept, namely the individual's social identity. Social identity is about what the individual feels s/he has in common with some people and is different from personal identity which refers to what differentiates the individual from others. Identities are often treated as fixed and unchanging. However, Jenkins (1996) suggests that social identity is also the resultant of how the individual thinks the others sees him or her. This makes identity construction the result of interaction. From the perspective which is adopted in this chapter, identities are constructed within discourses and practices and produced within specific historical and institutional contexts (Fay 1996). This perspective emphasizes the dynamic process of identity construction or identification. Identity construction involves the extraction of cues; it is social, retrospective, and ongoing. In this observation several of the characteristics that Weick (1995) sees as central to sense-making can be recognized (Pratt 1998, p. 180).

The recent attention for identity construction is based on the assumption that the construction of identities has become more problematic due to the fragmentation of society in the postmodern era. Because the individual interacts in a growing and sometimes fast-changing variety of groups the identity construction has become a 'reflexive project of connecting personal and social change' (Giddens 1991, p. 33). Identity construction is a never-ending process of differentiation and exclusion, more than it is of unity. By stating who you belong to you also make a statement on who you are not; therefore the social identity must be powerful, or it will not be gratifying (cf. Bauman 1999). According to Van der Loo and van Reijen (1997) it is the paradox of individualization that it has given the individual the freedom to choose, but at the same time the individual is, more than ever, forced to choose. The process of inclusion and exclusion makes the choice whether or not to join a group a political one. Thus, sense-making may be needed in order to determine which group one should identify with, within the possibilities and restraints that the context offers.

Authors disagree on whether under the condition of a fragmented society it is still possible for the individual to build a coherent identity. According to Giddens (1991) fragmentation does not automatically lead towards multiple selves and it may just as well lead to an integrated identity. Bauman (1995), however, suggests that due to the fragmentation of postmodern society the individual can no longer give meaning to his or her experiences and, therefore, can no longer build a coherent self. In fact, he states that 'the *postmodern* "problem of identity" is primarily how to avoid fixation and keep the options open' (Bauman 1996, p. 18). Although in this chapter the focus is on the construction of what appears to be a single identity namely the identity of graphic designer, it follows Hall's (1996, p. 4) suggestion that identities are constructed in a discourse in which multiple, sometimes conflicting identities are of influence. The assumption in this chapter is that within a specific historical and institutional context multiple and sometimes conflicting identities lay the foundations for identification, and that identification is the process in which new identities may arise.

This still leaves the question whether identification and commitment are different names for the same concept. The literature on the concept of organizational commitment shows the same confusion regarding the concept of identification. Most concepts of organizational commitment that are developed focus on the attitudes on which commitments can be based. An often used and refined model of attitudinal organizational commitment is that of Allen and Meyer (1990) in which three components can be distinguished, namely, the affective, continuance and normative component. The usefulness of this model is that it recognizes that commitment can be the outcome of affective, calculative and institutional motives. In this chapter the affective component refers to the individual's emotional attachment to an organization, based on identification. The calculative component refers to the individual's perception of the costs and risks associated with leaving their organization – the continuance component of Allen and Meyer – and to the individual's perception of a fair deal, which is often referred to as the calculative component. Finally, the normative component refers to a person's felt obligation towards the organization based on, for example, a work ethos.

Although most literature focuses on organizational commitment, it is commonly recognized that the organization is not an integrated entity and that people can be committed to various aspects of the organization, for example, their job, colleagues or the management. Furthermore, it is assumed in this chapter that in the dynamic process of identity construction commitments to other objects or groups outside the organization may be of influence.

This chapter perceives commitment as different from identification; commitment is defined as an attitude that is a possible outcome of identification with the object or an outcome of calculative or normative reasons. Behaviors that could be the result of commitment are a willingness to show good citizenship and/or the intention to stay member of the group. Whether these behaviors will actually be shown depends on the influence of various organizational, personal and contextual factors (Arnold, Cooper and Robertson 1998). The focus on attitudinal commitment does not exclude the existence of other forms of commitment, for example, commitment based on a formal contract or behavioral commitment.

This conceptual framework can be used heuristically in the study of sense-making by creative professionals.The various aspects of the concept commitment can be helpful as sensitizing concepts to elicit and make sense of the meaning creative professionals give to commitment.They suggest ways of exploring their identity construction and commitments in relation to, for instance, the meanings which they met in the art colleges which they visited and the sense these made of the identity of creative professionals and the reputation of specific graphic-communication agencies.

TRADITIONAL AND NEW PROFESSIONALS AND COMMITMENTS

There may not be much theoretical consensus on the concepts of commitment and identification; there is not much consensus on the concept of the professional either. Traditionally the concept of professional is associated with the medical and legal professions (Abbott 1988). While professions such as law, architecture or accountancy have norms and systems of conduct which are established by the state and their own institutional bodies, creative design does not have such normative systems (Julier 2000, p. 34). In the creative profession one can recognize characteristics of the new professional that Auer (1994) distinguishes. Creative professionals perform high-grade tasks, they work autonomously, their assignments are determined in interaction with the customer and most of the designers are not organized in a professional community. According to Gastelaars (1997), professionals have the tendency to attach great importance to their autonomy and professional responsibility which may lead to an, often unconscious, higher loyalty to their own profession than to the organization they work for. Research among creative professionals (Leisink 1995; van Wijk 2000) shows that they share in these professional characteristics.

Bruel and Colsen (1998) suggest a dualistic distinction between two types of professionals. Professionals are either committed to their profession or committed to the organization. In this distinction the differentiation between 'locals' and 'cosmopolitans' of Gouldner can be recognized (Slootman 1991). The 'local' professional has an affective commitment with the organization, whilst the commitment of the 'cosmopolitan' professional is focused on external cues such as the profession. The relation between the organization and the 'cosmopolitan' professional is based on calculation as far as the professional is concerned, which implies that if the employment deal loses its appeal the professional will move on to a better deal.

Other researchers have questioned this dualism and suggest a more dialectic approach. Schomaker's (1999) research of accountants, for example, shows that in a professional organizational structure organizational commitment can coexist with professional commitment. Nevertheless, she also shows that accountants may decide to look for new employment despite their organizational identification. In her conclusion Schomaker suggests that this decision is a result of commitment to the profession.

If we combine the possibility that professionals may have coexisting commitments together with the theoretical perspective on commitments that was elaborated in the previous section, the thesis can be formulated that more commitments could be of influence in the decision-making process of creative professionals whether or not to stay in the organization, and that this process may be more dynamic than research has recognized so far. This calls for an awareness of other situations that could be perceived of influence on the decision-making process such as the influence of the private life (cf. Hochschild 1997), the influence of personal needs (cf. de Moor 1992), the influence of organizational strategy (cf. Sennett 1998), the perception of a fair deal (cf. Guest 1998, quoted in Koopman et al. 2000), the economic situation or the relatively young age of creative professionals (Hartog, Mekkelholt and van Ophem 1988). In the following sections we will examine this thesis on the basis of empirical data.

IDENTITY CONSTRUCTION AND COMMITMENTS OF CREATIVE PROFESSIONALS: A CASE STUDY

Below, a case study is presented of identity construction and commitments of creative professionals. First a description is given of the method of research and the research context. The career path of the creative professionals will serve as a guideline in the description of the process of the construction of the identity of the graphic designers and their commitments. The influence of art college on the construction of the identity of graphic designers will be given attention. Then the influence of work experiences on the decision whether or not to become a freelancer is discussed. Finally, the commitments of the graphic designers and their influence on the decision-making process are examined.

Method of Research

Adopting a constructivist perspective implies that the view which creative professionals construct in an open interview is different from the view which researchers construct when they invite creative professionals to react to the constructs that they submit to them in, for example, a survey questionnaire. To do justice to this implication, this chapter is based on the outcomes of open interviews held with eight graphic designers in a design company and one designer who had left the company to become a freelancer in the autumn of 2001. The graphic designers were asked to relate the biography of their educational and working careers, and what they perceived to be of influence on the choices they made. The outcomes can be placed in a broader context by comparing them with the outcomes of open interviews that were held with other professionals of the company such as project managers and web editors. In addition to open interviews, observations took place on the shop floor and in performance and job interviews held by the agency management. It is worth noting that the data cannot be generalized to other populations because of the limited size of the research population and the specificity of the context.

The Research Context

The research was conducted in a design studio referred to in this chapter as One of a Kind, which has its residence in the centre of the Netherlands. The developments in the sector as described in the introduction can be recognized in One of a Kind's extension of services from graphic design towards graphic communication and the introduction of web design. Furthermore, One of a Kind operates in a network of communication and advertising agencies and has a mix of local and internationally orientated clients.

Since its foundation by two college friends, the organization has grown into a company with 24 employees, including eight designers, which is representative of most advertising agencies, although rather large compared to other design companies (van Wijk 2000). One of a Kind can be characterized as a professional organization with a simple structure that is managed top down; the two founders define the strategy, give art direction, manage the office and maintain contact with the clients.

The company is established, as most agencies in the branch, in a trendy office, built in an old factory in the centre of town. Behind the front office situated in the entrance hall, seven graphic designers share an almost open workspace on the ground floor, together with three database publishers and three web editors. Two graphic designers specialize in graphic design, and three designers in Internet design, two designers specialize in Internet as well as graphic design. This division used to be visual on the work floor as well. A glass wall separated the Internet designers from the rest, but due to a shortage of workspace the various functions are now combined on the work floor. The atmosphere on the ground floor is lively, most of the time music is playing, formal and informal meetings take place and people move back and forth to the big kitchen in the back of the ground floor. The kitchen is the centre of the building where guests wait for their appointment, people take a break, informal client meetings are held and all join for lunch. The top floor is divided into four lined-up rooms and is the workplace of the management, the project managers and one graphic designer who prefers quieter surroundings.

One of a Kind has a young population in age as well as by seniority. The average age of the employees is 30, which is comparable to other advertising agencies (van Wijk 2000). Apart from one designer, all designers are still junior designers and for three of them One of a Kind is their first employer. Five out of the eight designers have been employed for less than a year at One of a Kind. The mobility rate is high; most creative professionals at One of a Kind leave the organization within two years to start as a freelancer. The high mobility rate is comparable to other advertising agencies (van Wijk 2000). Management accepts the mobility rate as a fact of life, but they perceive a lack of commitment, which refers in their opinion to good citizenship, of the employees as problematic. In return for all they offer such as inspiring accounts, free lunches, massages and outings, management expects their employees to take full responsibility for their job, including responsibility for the planning of time and money. In their opinion the employees lack this responsibility especially at a time of economic recession. The management, however, was also willing to consider itself as a possible cause of the lack of employee commitment and responsibility, and hired a consultancy agency to examine possible improvements within the management. This investigation partly overlapped the research period.

Generated by the expanding Internet market and several awards they received for their designs, business was flourishing for One of a Kind until the summer of 2001. During the time of the research the effects of the collapse of the Internet market and the terrorist attack in America became visible. However, although business was slow One of a Kind was not forced to cut down on staff, unlike other agencies in the sector. In fact, the organization still had vacancies for outstanding senior designers that turned out to be difficult to fill. According to the management this is caused by the tendency for senior designers to start their own business and the fact that most designers are focused on Amsterdam as the capital of the design sector.

The Construction of the Identity of Graphic Designers at Art College

A general characteristic of the designers of One of a Kind is that they received their education at an art college. They all perceived the step to go to art college as natural. Since their childhood they have always been drawing pictures and using this creativity appeared a logical outcome, although the path to follow varied per individual. For some designers, going to college has always been the goal; others first started a completely different course before they discovered they preferred to use their creative capacity. This creative characteristic distinguishes the designers from other professionals in the organization. One project manager stated:

> [The graphic designers], of course, have always been creative ... and then you visit graphical college or go straight on to art college. It is a straight line ... But I have the feeling I swing from one end to the other before I reach my goal ... for example, in my previous job I found the right branch and in this job I have found the right position as well.

At most art colleges the first year is a general year in which the students are taught to lose their old habits in order to find their own style in design. After this year the students choose their specialization. Except for the designers who first attended graphical college, most designers at One of a Kind only decided to specialize in graphic design in this year. Although the profession of graphic designer is not legally protected, all creative professionals at One of a Kind, including the founders, share the opinion that in order to become a graphic designer an education at an art college is required, or as one of the directors stated: 'A graphical college can teach you the technique but to become a designer you need to go to art college'. Graduation from art college appears to be a powerful resource of inclusion and exclusion at One of a Kind. One of the web editors has the ambition to become a graphic designer for which he feels his graduation at graphical college is sufficient. He knows, however, that he will never get the chance at One of a Kind to develop graphic designer skills without a graduation at art college.

Although all designers received their education at an art college, this education is not perceived as equal. Some art colleges have a better reputation than others and so have their graduates. This reputation might be linked to the degree of independence that is taught at art college. The Rietveld Academy, for example, has a good reputation and is known for the emphasis they put on independence in art and design.

The need for artistic freedom appears to be a strong characteristic of the identity of graphic designers. When asked to motivate their decision to become a graphic designer, several of the designers first explained why they did not choose to become artists before they motivated their decision to become a graphic designer. It seems that the free artist is held in higher esteem than a graphic designer, because graphic design is perceived as a more applied art form. Even though for some designers the fear of total freedom or the need for financial security has been a reason to specialize in graphic design, all designers give the same answer to the question what would be a reason for leaving One of a Kind; 'the loss of creative freedom'. The wish to become a freelancer in the future, which many graphic designers expressed in the interviews, seems to be based on this need for creative freedom.

There are some indications that this need first started to develop at art college. One of the designers said that he is struggling with the fact that he feels comfortable working at One of a Kind, although art college taught him to become an independent designer. Another designer mentioned in the interview that he chose to attend graphical college first, because he was afraid art college would be too vague and once he entered art college he felt frustrated because he had to let go of everything he had learned before. In his final year, however, he and some friends decided to break free from the unwritten rule that graduates of their college end up working in big advertising agencies, and started their own business.

The identity of graphic designer appears to have two strong characteristics, namely, the need to make professional, creative products and the value of creative freedom. These examples show the influence of art college on the construction of the identity of the graphic designer in two ways: (1) by taking a specialized education at art college you become a member of the group of professional graphic designers; and (2) the value of creative freedom appears to be constructed at art college. Julier's analysis (2000) of the culture of design provides several insights that help to understand how art schools perform a role in this. Creative potential is the driving principle in art schools, in the selection of new students and the (lack of) organization of the curriculum. The emphasis is on individual creativity. Because of this, art schools have always resisted state attempts to make art and design education more vocational and compliant with formal regulations which prevail in the educational system. According to Julier (2000, p. 36) it is also the 'myth' of individuated creativity inherited from art education, on which designers draw to differentiate themselves from each other. Of course, creating a distinctive creative profile is also a means of competing with each other for clients' interest (Julier 2000, p. 37). Thus individual creativity and absence of formal procedures are established from the beginning as inherent features of the social identity of creative professionals. In the Netherlands, the value of creative freedom of designers has become institutionalized in the *starters stipendium.* The starters stipendium is a financial gift that is granted to promising business plans, and is meant to give designers who start their own business in their first year the financial freedom to develop their skills further and build a fruitful network. Designers can apply for this stipendium, within two and a half years of graduation.

Although art college seems to have a major influence on the identity construction of the graphic designer, this influence is not deterministic. Our findings show that in the end the designers make their own decisions about their careers. After graduation the creative professional has the option of becoming a freelancer or an employee. Becoming a freelancer appears to fit the value of creative freedom best. However, all designers in this study chose the option of employment, although two designers combine this employment with a career as a freelancer. In the next section we will examine which factors influenced the graphic designers' decision to accept employment at One of a Kind, and which cues are of influence in the decision-making process regarding future careers.

Graphic Designers at One of a Kind

Various characteristics were mentioned by the graphic designers as reasons why they had applied for a job at One of a Kind. All designers said that they could identify themselves with the work of One of a Kind. According to them characteristics of the work are that it is challenging, diverse and fun. Another attractive aspect of One of a Kind is the interesting accounts. Furthermore, One of a Kind is considered a good learning environment. Over the years One of a Kind has won several awards with graphical as well as web designs, which is why they have become famous in the designers circuit, and most designers stated they felt that they could learn the ropes here. Finally, a few designers mentioned atmosphere and colleagues as reasons why they had applied for a job at One of a Kind.

When asked whether they still enjoyed working at One of a Kind after some time all designers said they did. Some designers said a dream came true when they were hired by One of a Kind. They felt admiration for the management, because they admired the products they made over the years. In time, however, a shift could be observed in the importance of these marks. Once the graphic designers have created their own products, admiration for the management changes into a feeling of equality. Owing to this shift, designers perceive fewer learning opportunities. At this point most people get restless in their job and start to wonder about their position in the organization. Most designers mentioned in the interview that they were considering whether their future career should be at One of a Kind or whether they should work as a freelancer. The perception of the amount of creative freedom that One of a Kind offers seemed to be one of the main considerations in the question whether to stay or become a freelancer. At this point diversity could be observed in the way designers give meaning to the organization's characteristics that play a role in the decision-making process.

Although all designers identify themselves with the work of One of a Kind, not all experience the same creative freedom in their designs. Some perceive the art direction given by the management as a useful brainstorm that complements their own design and stimulates their personal development, whilst others sometimes feel restricted by the 'design gravy' that the art directors add to their design. When asked why being a freelancer appears to be so appealing many designers answered that being a freelancer means you have 'total control over the design'. Finally, an interesting point is that, in the perception of some designers, freelancing offers the creative freedom to explore interests outside graphic design like film, music or photography; interests they do not find the time for now. These graphic designers said they perceived themselves more as designers than as graphic designers.

The search for creative freedom might also be the reason why none of the designers considers applying for employment at another company, although the reputation of One of a Kind might also be of influence. According to some designers their goal was to work for a well-known office. Working at One of a Kind has made that dream come true and they feel there is no use in trying to work at another office, because there is nothing more to achieve.

The reputation of One of a Kind may also have another effect on the decision-making process. All designers feel that as a freelancer it will be difficult to become recognized as an outstanding designer, while in the perception of some designers being employed in a famous organization creates the opportunity to become well known, although some designers fear that being employed at One of a Kind will imply they always have to stand in the shadow of the owners. However, some designers said fame was less important than creative freedom.

A perceived positive side-effect of reputation is that it attracts interesting accounts. Once you have built a reputation clients will choose you for your work, which generates more creative freedom. All designers identify with the statement of One of a Kind that they know what is best for the client although this may not always be the solution the client perceives as best, but admit that it will be difficult to hold on to this statement when starting as a freelancer. For some designers this is a reason to stay employed at One of a Kind, although some designers give a more practical reason, which is a perceived lack of skills to attract interesting clients themselves.

The final characteristic that was mentioned is the atmosphere of the organization. Although only a few designers mentioned atmosphere as an important reason for wanting to work at One of a Kind, all designers said it was an important cue in their decision-making. The designers use the word atmosphere in two different ways. The first way refers to the informal style in which the work process is organized. All designers state they would leave the organization if it turns into a bureaucratic office such as, as they call it, 'a big advertising agency', in which processes are standardized too much and the client has too much power over the outcome of the design. To them, this type of organization represents the loss of creative freedom.

The other meaning of atmosphere refers to the contact with other colleagues. All designers appreciate the informal contacts on the work floor and some designers mention friendships with other colleagues. However, the need for creative freedom has its impact on the interaction with colleagues. In an art direction meeting with all designers it was mentioned that the creative process and its outcome are perceived as personal and an expression of the ego, in which no interference is allowed and cooperation with other designers is perceived as problematic most of the time unless the designer feels he or she may benefit from this cooperation. In general, however, designers appreciate the value of colleagues as sparring partners. Some designers would miss this feedback and hesitate to become freelancers, whilst others, if necessary, would seek contact with other freelancers to fulfil the need for a sparring partner.

Although it is not the intention of this chapter to give a complete overview of possible influences on the decision-making process of creative professionals some other cues were mentioned that seem worthy of being discussed here as well. As mentioned before, some designers based their choice of becoming a graphic designer on their need for financial security. This same need appears to be of importance in the decision-making process whether or not to become a freelancer. Two designers decided to combine a career as a freelancer with employment at One of a Kind, which in their opinion is the ideal combination of artistic freedom and financial security, until their own business has lifted off. Others mentioned the economic setback as a reason for postponing the step to freelancing. Although there is a recession, some designers perceive the starters stipendium as an attractive impulse to start their own business, and because time is running out for them they have to make a decision soon. One designer recently received a starters stipendium, and left One of a Kind to start her own business.

Some designers said the repertoire of their function was of influence in the process of decision-making. Owing to task specialization designers spend a great deal of their time behind the computer. This may not only cause repetitive strain injury (RSI) problems from which some designers already suffer, some designers state that their job has become less attractive while they miss the contact with clients and printing-offices. For some these might be reasons to become a freelancer or to develop other interests, outside graphic design.

The last cue mentioned here is the influence of the private situation. Some designers said that their need for spare time influenced their decision-making. Being a freelancer asks for an investment in time. Some designers mentioned the importance of spare time to do other things or to spend time with their partner as a consideration in the decision-making process.

These examples show that the decision to apply for a job at One of a Kind seems grounded in three major motives: the need for security, the need to develop as a designer and identification with the image of the organization as perceived by the designers, to be more specific with the products, the learning environment, the accounts and the informal structure. This identification seems to be based more on a perceived similarity between the identity of graphic designer and the image of the organization than on an internalization of the values and norms of the organization.

In the early stage of their employment the perceived possibilities for personal development appear to be the strongest cue to stay employed. After some time the designers perceive their position in the organization as more equal and the learning aspect disappears as a characteristic of the image of the organization as perceived by the designers and the need for creative freedom, and therefore the wish to start a career as a freelancer appears to become stronger. A major theme in the decision-making process whether or not to become a freelancer seems to be the question whether the image of the organization as perceived by the designers is still similar to the perceived identity of graphic designer, especially when it comes to creative freedom. Although presented here as separate, a mixture of cues appears to have its influence on this process. The designers sometimes indicate that similar cues are important; however, they base their individual decisions whether to stay at One of a Kind or become a freelancer on the meaning they attach to these social, organizational and personal cues, perceiving them, on the one hand, as possibilities and, on the other hand, as barriers.

In this process of sense-making a conflict between the individual perceptions of the identity of graphic designer, the image of the organization and other personal identities can be recognized. The following section will take a closer look at how the designers make sense of this conflict by exploring whether or not the commitments the designers perceive are of influence on the decision-making process on future careers.

Graphic Designers and their Commitments

Grounded on sayings such as: 'If I lose my creative freedom I will leave', or, 'When my personal development stops I will be gone', it might be suggested that the decision whether or not to stay in the organization is the outcome of calculative reasons that are based on the need for creative freedom and personal development. These needs could be conceptualized as professional commitments. However, the designers identify themselves with several characteristics of the identity of One of a Kind. The question arises whether this identification with the organization may lead the graphic designers to feel committed to the organization as well.

Five out of the eight designers have been employed for less than a year at One of a Kind. When asked about their future plans most designers state that they intend to stay in the organization for a maximum of one or two years. In a group discussion, however, indications for latent organizational commitment as an outcome of identification were found, which is suppressed for various reasons. Some designers confessed they find it difficult to admit that they enjoy working at One of a Kind and think about a future inside the organization. According to them this hesitation follows from the influence of art college, but also from the international arena. There are so many possibilities out there, that it feels a waste of talent to stay in this organization; some even stated they feel pushed by this feeling. Furthermore, in an informal meeting one designer confessed he is afraid to express his empathy towards the organization because he has the feeling the management might punish him for this in salary discussions. To gain a higher salary profit, he feels it necessary to present the option of leaving the organization as realistic, although this same designer expresses thoughts about a future role for himself in the management team.

Another designer has the wish to stay part of the organization and this wish seems an outcome of the fact he does not want to become a freelancer. Although discussions are going on about him joining the management team, he still has his doubts. He prefers to become 'a cell within the organization', so he can profit from all the benefits of working at One of a Kind such as interesting accounts, project managers and a well-equipped work space, without having the responsibilities of a manager that in his perception have nothing to do with what he loves most, graphic design. Although calculative motives seem the main reason for him to stay in the organization, he also mentioned that the only organization in which he wants to be a cell is One of a Kind and that he feels committed to the management.

Although most of the designers have the intention to leave the organization and become a freelancer, in their perception this wish does not exclude organizational commitment. All designers perceive themselves as a member of One of a Kind, and think of themselves as committed employees; they find it normal to put in extra efforts to reach for the best, and they work extra time when needed, although they admit that they sometimes show a lack of responsibility when it concerns specifics, such as financial or efficiency matters, which seem to lack as a characteristic of the identity of graphic designer. Organizational commitment therefore seems closely related to professional commitment. This could explain why in the perception of the management the designers lack organizational commitment.

It is worth noting that in the perception of some designers the meaning of commitment does not refer to their attachment to the organization but to the efforts the management makes to involve the employees in the organization. Some designers complain that the management asks for extra efforts of the employees, while they do not show commitment to them. Due to this lack of commitment of the management, some designers feel reluctant to put in extra effort, something which Torka and van Riemsdijk (2002) also found in their study of mechanics and welders.

Apart from commitment to the profession and the organization, the designers also feel committed to their colleagues. The need for professional feedback of colleagues was mentioned before, but in addition colleagues are perceived as important for social well-being. One of a Kind has to feel like home, but at the same time it is perceived as a professional workplace. The designers feel there should be a balance between social aspects and the profession. If one of the two disappears it is time to move on. For some time, the social contact with colleagues can compensate for the lack of interesting work, and vice versa. However, too much social interaction disturbs the concentration on work, just as working too hard disturbs the feeling of well-being. The way in which creative professionals make sense of this type of commitment exemplifies the dual meaning of 'organization' in which relationships are both part of the production process with its technical rationality and part of the 'community' with its social rationality (Verweel 2000). But, notwithstanding their commitment to the organization and the importance of the social aspect for the designers, in the end they will always show more commitment to their profession, or as one of the designers put it: 'At least four people that have left since I work here, have become a freelancer … The funny thing is that people at One of a Kind are very close but despite this people leave the organization after a couple of years ... But the truth is they all leave in great distress.'

The data show that the need for creative freedom does not exclude the possibility that organizational commitment and professional commitment coexist, together with a commitment to colleagues. Professional commitment seems based on a strong identification with the identity of graphic designer, but organizational commitment and commitment to colleagues may be based on calculative reasons as well as on identification. It does seem that neither of these commitments is useful as an indicator of the intention to stay in the organization. It appears that some designers feel organizational commitment in the sense of retention, but they suppress this for various reasons, for example, the wish for a higher salary profit.

Organizational commitment in the sense of retention appears to be situational and dynamic, but so is professional commitment in the sense of freelancing. As shown in the previous section, various cues are of influence in this decision-making process. The above confirms the assumption that the process of identity construction is more dynamic than research has recognized so far.

In the previous section indications were found of an ambiguity between the identity of graphic designer and the identity of the organization. The difficulty to express their commitment to the organization that some designers experience shows that there might not only be a conflict between the identity of graphic designer and the identity of the organization, but also between being an employee and being a graphic designer. For most designers the solution to these conflicts is to leave the organization and become a freelancer, as the mobility rate of employees of One of a Kind shows. Nevertheless, for some designers leaving the organization is not an option. Some designers wish to stay members of the organization because they perceive a strong affection for the organization. Other designers feel forced to stay in the organization because they perceive a lack of possibilities to become a freelancer. It might be that the designers who feel themselves forced to stay in the organization will have more problems making sense of the conflicts mentioned, although further research should be conducted on this matter.

CONCLUSIONS AND SOME POINTS THAT DESERVE FURTHER EXAMINATION

The objective of this chapter was to explore the meaning creative professionals give to commitment and the origins of these commitments by exploring the construction of the identity of graphic designer.

The process of identity construction has a recursive character as appears from the fact that creative professionals perceive their creativity as a personal identity characteristic on which they base their decision to go to art college, while especially art college and, in the case of some designers, the organization appear subsequently to be of influence on the construction of the social identity of graphic designer. This influence is not deterministic in the sense of occurring without the agency of the (prospective) designer. The data showed that the designers enact their individual decisions based on the meaning they attach to social, organizational and personal cues. Furthermore, indications were found that various conflicting identities appear to play a role at some stages in the process of the construction of the identity of graphic designer. A conflict was found between the need for creative freedom, one of the main characteristics of the social identity of graphic designer, and the identity of employee of One of a Kind. In the process of making sense of this conflict the perceived identity of the organization, other social identities, personal characteristics and contextual factors like the economic recession appear to have an influence, thereby illustrating the assumption that within a specific historical and institutional context multiple and sometimes conflicting identities lay the foundations for identification, and that identification is the process in which new identities may arise. Depending on the perceived constraints and possibilities, the solution to this ambiguity that is most favoured by the designers is to actually become a freelancer. In this process the paradox of individualization can be recognized. A great deal of opportunities arise, but at the same time the creative professional is forced to make a choice. Indeed, some creative professionals mentioned that they appreciate the freedom to choose, but sometimes feel insecure as to whether they make the right choices.

In the discussion about the relation of the concepts of commitment and identification, commitment was conceptualized as an attitude that is a possible outcome of identification or of calculative or normative reasons. The data appear to subscribe to this differentiation between identification and commitment; although the creative professionals identify themselves with organizational characteristics or colleagues, these identifications not always lead to commitment and indications were found that designers show various commitments based on calculative motives. Whether or not identification leads to commitment or commitment is based on identification depends on individual sense-making. Furthermore, confirmation was found for the thesis that the identification process may result in a broader variety of commitments than just professional commitment or organizational commitment. Indeed, the outcomes show that in the perception of creative professionals, professional commitment can coexist together with organizational commitment and commitment to colleagues, although professional commitment appears to be the strongest.

Even more striking in the research findings is that neither organizational commitment, nor commitment to colleagues, nor professional commitment appear to be useful as indicators of the intention to stay in the organization. The decision-making process whether or not to become a freelancer appears to be more dynamic than research has recognized so far. Furthermore, indications were found that the intent to stay is not only an outcome of commitment but is also influenced by the perception of the market. These outcomes do not imply that the concept of commitment has become a useless concept. It seems useless as an indicator for the intention to stay in the organization, but it appears to remain a useful concept for good citizenship in the sense of exertion. The designers are willing to put in extra efforts to reach for the best, although their definition of exertion is a subjective one that might not fit that of the management.

One might object to the findings of this study that future research may reveal that the graphic designers in this study actually miss the capacity to become a freelancer and that their wish to leave the organization in order to become a freelancer is wishful thinking. Another objection to the findings could be that the designers do not threaten to leave the organization with the intention of actually becoming a freelancer, but as an attempt to change the situation inside the organization, as Hirschman (1970) suggests. According to Hirschman loyalty to the organization activates the voice option and the effectiveness of the voice mechanism appears to be strengthened by the possibility of exit. Although the example of the designer who threatened to leave the organization but in fact wanted to gain a higher salary profit confirms the latter, the fact that all of the designers who have left the organization so far have become freelancers seems to confirm that the intention of becoming a freelancer is not merely voiced as a threat and a means to other goals.

Although previous surveys held among employees in advertising agencies (Leisink and Spaninks 1993; van Wijk 2000), appear to confirm the findings of this study, the size and specificity of the research population in the case study prevent drawing general conclusions about the construction of identities and commitments. Many questions are still left unsolved. Due to the fact that the case study covers only a relatively short period of time, it was difficult to examine, for example, the influence of the relatively young age of the respondents and that of the economic situation – a factor which Hartog, Mekkelholt and van Ophem (1988) found to be of influence in their study on mobility of employees. Furthermore, in this study indications were found that the reputation, product, management strategies and structure of organization could have an effect on the identity construction and commitments of creative professionals. Therefore, future research on the influence of various sets of cues on identification processes would stand to benefit from including other types of organizations.

NOTE

1. The author wishes to thank Willem Koot and Peter Leisink for their comments and suggestions for improvement.

REFERENCES

Abbott, Andrew (1988), *The System of Professions: An Essay on the Division of Expert Labor*, London, UK and Chicago, US: University of Chicago Press.

Allen, N. J. and J. P. Meyer (1990), 'The measurement and antecedents of affective, continuance and normative commitment to the organization', *Journal of Occupational Psychology*, **63** (1), 1–18.

Arnold, John, Cary L. Cooper and Ivan T. Robertson (1998), *Work Psychology: Understanding Human Behaviour in the Workplace*, Harlow: Pearson Education.

Auer, C. (1994), 'Leidinggeven aan professionals. Lessen van kannibalen, kunstenaars en kinderen', *Management en Organisatie*, **3**, 301–26.

Bauman, Zygmunt (1995), *Life in Fragments: Essays in Postmodern Morality*, Oxford: Blackwell.

Bauman, Zygmunt (1996), 'From pilgrim to tourist', in Stuart Hall and Paul du Gay (eds), *Questions of Cultural Identity*, London, UK and Thousand Oaks, US: Sage, pp. 18–36.

Bauman, Zygmunt (1999), *Culture as Praxis*, London, UK and Thousand Oaks, US: Sage.

Bruel, Maurits and Clemens Colsen (1998), *De geluksfabriek: Over het binden en boeien van mensen in organisaties*, Schiedam: Scriptum Books.

Fay, Brian (1996), *Contemporary Philosophy of Social Science*, Oxford, UK and Malden, US: Blackwell.

Gastelaars, Marja (1997), *Human Service in veelvoud: Een typologie van dienstverlenende organisaties*, Utrecht: B.V. Uitgeverij SWP.

Giddens, Anthony (1991), *Modernity and Self-identity*, Cambridge: Polity Press.

Hall, Stuart (1996), 'Who need "identity"?', in Stuart Hall and Paul du Gay (eds), *Questions of Cultural Identity*, London, UK and Thousand Oaks, US: Sage, pp. 1–17.

Hartog, J., E. Mekkelholt and H. van Ophem (1988), 'Arbeidsmobiliteit in Nederland 1957–1985', *Tijdschrift voor Arbeidsvraagstukken*, **3**, 65–74.

Hirschman, Albert O. (1970), *Exit, Voice and Loyalty*, London, UK and Cambridge, US: Harvard University Press.

Hochschild, Arlie R. (1997), *The Time Bind: When Work Becomes Home and Home Becomes Work*, New York: Metropolitan Books.

Jenkins, Richard (1996), *Social Identity*, London, UK and New York, US: Routledge.

Julier, Guy (2000), *The Culture of Design*, London, UK and Thousand Oaks, US: Sage.

Koopman, Paul L., Nanja J. Kolk, Frank van Luijk and Jaap J. van Muijen (eds) (2000), *Boven het maaiveld*, Amsterdam: Uitgeverij Nieuwerzijds.

Leisink, Peter (1995), 'Autonomie en nieuwe arbeidsverhoudingen in de zakelijke dienstverlening', in Jules J. J. van Dijck, Erik Henderickx and Jacques J. B. M. van Hoof (eds), *Baas over eigen loopbaan*, Houten: Educatieve Partners, pp. 62–76.

Leisink, Peter and Louis Spaninks (1993), *Human Resource Management en Arbeidsbeleving in de Reclamebranche*, Amsterdam: VEA.

Moor, Wim de (1992), *Arbeidsmotivatie als management-instrument*, Houten/Diegem: Bohn Stafleu van Loghum.

Mowday, Richard T., Lyman W. Porter and Richard M. Steers (1982), *Employee-Organization Linkages: The Psychology of Commitment, Absenteeism, and Turnover*, New York: Academic Press.

Pratt, Michael G. (1998), 'To be or not to be? Central questions in organizational identification', in David A. Whetten and Paul C. Godfrey (eds), *Identity in Organizations: Building Theory through*

Conversations, London, UK and Thousand Oaks, US: Sage, pp. 171–207.

Schomaker, Pieternel A. (1999), '*Wie bindt, die wint*', Nijmegen: Nijmegen School of Management.

Sennett, Richard (1998), *The Corrosion of Character*, London, UK and New York, US: Norton.

Slootman, Arnold W. (1991), *Arbeidsbeleving van universitair wetenschappelijk personeel*, Utrecht: Lemma BV.

Susijn, Wouter (1999), *HRM: luxe of noodzaak?*, Amsterdam: VEA.

Tajfel, H. (1981), *Human Groups and Social Categories: Studies in Social Psychology*, New York: Cambridge University Press.

Torka, N. and M. van Riemsdijk (2002), 'Betrokkenheid en werkgeversgedrag', *Tijdschrift voor HRM*, **4**, 35–58.

Van der Loo, Hans and Willem van Reijen (1997), *Paradoxen van modernisering*, Bussum: Coutinho.

Vermeulen, Jeroen (2001), *De naam van de school*, Baarn: Ten Have.

Verweel, Paul (2000), *Betekenisgeving in organisatiestudies*, Utrecht: Universiteit Utrecht.

Weick, Karl E. (1995), *Sensemaking in Organizations*, London, UK and Thousand Oaks, US: Sage.

Whetten, David A. and Paul C. Godfrey (eds) (1998), *Identity in Organizations: Building Theory through Conversations*, London, UK and Thousand Oaks, US: Sage.

Wijk, Ellen van (2000), *Arbeidsbeleving bij communicatie-adviesbureaus 2000*, Amsterdam: VEA.

PART III

Shifting Inter-Organizational Relationships and Identity Formation

10. The Construction of Identity in a Fragmenting Telecom Operator

Alfons van Marrewijk

INTRODUCTION

Recently I received a letter from the Dutch telecom service provider, KPN, announcing the sell-off of its International Consultancy Unit. This was the division I was working in part-time. Faced with an astronomical debt, KPN had decided to focus on core business, which would not include international consultancy services. What should I do? Apply for a job at another consultancy company or within KPN? How far had I identified with KPN? Did I see myself as an international consultant or as a KPN employee? These questions of identification have recently haunted me and my (ex-)KPN colleagues.

This identity crisis is expressive of the growing importance of identity construction in modern organizations. The construction of individual identity has become a reflexive process in which one must find one's identity amid the options offered by abstract systems in society (Giddens 1990). As modern organizations are being increasingly organized in web-like structures of small companies and strategic alliances, the questions of 'who we are' and 'what we identify with' have become important (Castells 1997; 2000). This trend is also discernible in the telecom sector.

New technological developments, new government policy on telecommunication, and the growing awareness that telecommunication is a key factor in the international trade system have brought fundamental changes to the telecom sector (van Marrewijk 1997; 1999). Only 15 years ago, telecom operators were government-owned monopolistic bureaucracies. Now they are organized in networks of strategic alliances and partnerships. To what extent have these fundamental changes in the structure of telecom operators affected the way in which their employees construct their social identities?

Corporate identification is understood as a specific type of social identification which occurs when an individual uses his or her conception of the organization to define his or her identity (Pratt 2001, p. 14). If the identification of an individual is dependent on the organizational context

(Bartel and Dutton 2001), complex and cluttered organizational networks should result in a more difficult process of identification.

The aim of this chapter is to shed light on the fundamental changes in the structure of public-sector telecom operators and the way their employees construct their identity. The restructuring of Dutch telecom operator, KPN, will be explored for the detailed case description. The case of KPN is illustrative for other European public telecom operators. This chapter is structured as follows. First, it will explore the process of restructuring Dutch KPN from 1989 till 2001. It will then discuss the consequences of this process for the social identity construction of the KPN employees. Finally, the strategic Unisource alliance will be presented as an example of identity construction in an inter-organizational context. But first we need to take a closer look at identity processes in organizational contexts.

THE CONSTRUCTION OF IDENTITY IN ORGANIZATIONAL CONTEXTS

The social identity theory of Tajfel (1981) can be useful in studying the individual relationship with the organizational collective. This theory is based on the idea that the social category that the individual relates to provides a self-definition of who he is (Tajfel 1981). These social identities are not only descriptive and prescriptive, but also evaluative, based on subjective belief structures (Hogg and Terry 2001, p. 4). This last concept refers to the individual's ideas on the relation between his own group and other groups. Identity is the result of the definition of similarities and differences with other persons and groups, and is therefore open to subjective construction (Jenkins 1997). Invented ideologies and historical and ethnic schemes can be used to construct the relationship of individuals to their own group (Friedman 1994).

Organizational identification is a specific type of social identification; it is enhanced in traditional organizations when three conditions are present (Pratt 2001, p. 17). First, the organization must be distinctive so that it is more easily separated from other organizations. Second, the perceived differences between one's own organization and those in the external environment must be salient. This allows individuals to identify more easily with their own organization. Third, inter-organizational competition must exist between the individual's own organization and the other organizations. This minimizes the differences within his own organization and maximizes the differences with the external environment.

Individual identity can be constructed from diverse social identities 'as a basket full of selves' (Cohen 1994, p. 11). This perspective is reflected in the concept of nested identities in organizational contexts (Ashforth and Johnson 2001). The different levels of nested identities are job, work group, department, division and organization. Ashforth and Johnson (2001) argue that identity in an organizational context is determined by the subjective importance of the identities and by the situational relevance. Ashforth and Johnson (2001, p. 33) distinguish higher-order (organization, division) and lower-order (job, work group) identities. Lower-order identities tend to be more salient, while organizations invest considerable energy in boosting awareness of higher-order identities. The nested identities are connected with formal social categories and are institutionalized in the structure of the organization. To include both formal and informal groups Ashforth and Johnson (2001, p. 44) introduce the concept of 'cross-cutting identities'. In contrast to nested identities, cross-cutting identities are connected with social categories that are either formal or informal. The formal cross-cutting identities overlap the nested identities (for example, committees), while the informal cross-cutting identities can be distinct from the nested identities (for example, friendship cliques). This concept of nested and cross-cutting identities can be useful for exploring social identity processes in traditional bureaucracies, but does this also apply in the case of a complex context of unstable organizational networks?

In organizational networks it is difficult to set the three fore-mentioned conditions for organizational identification. The first and second conditions are not present as it is not clear whether temporary workers, tele-workers, external consultants, volunteers, part-timers, internal consultants and ex-pats are part of the organizational network. The third condition, inter-organizational competition, is also absent as organizations are linked up in multi-directional networks. In an ambiguous and complex context, organizational membership is not always stable and clear. This can result in the strategic use of identities by employees (Koot 1994). Bartel and Dutton (2001, p. 117) argue that processes of identification in network organizations are more difficult than in traditional bureaucracies. As the perception of organizational membership is context-dependent, individuals will find it difficult to define their identity in an ambiguous context. Bartel and Dutton (2001) use the term 'ambiguous' for situations in which membership is vague, problematic, and unstable. Ambiguity arises when individual membership status is unclear to the employee. When the sense of belonging to an organization is not triggered by the sharing of results of events between the employee and the organization it is difficult for the employee to evaluate his or her relationship with the organization and hence to develop a sense of belonging. The greater the ambiguity surrounding one's organizational memberships status, the more likely it is that a person will perceive him or herself as an outsider and experience feelings of insecurity and instability (Bartel and Dutton 2001, p. 119).

The exploration of the social identity theory in organizational contexts has resulted in the hypothesis that the process of constructing individual identity in the traditional bureaucratic telecom operators might have changed due to fundamental changes in the organizational context. It is necessary, therefore, to focus first on the process of change undergone by the Dutch telecom provider KPN to answer the question of whether KPN is still a traditional bureaucracy or a network organization?

THE CASE OF KPN: FROM BUREAUCRACY TO INTER-ORGANIZATIONAL NETWORK

In the General Agreement on Trade in Services (GATS) national governments committed themselves to restructuring their national telecom markets according to a specified time schedule. As a consequence, each government faced at national level the difficult task of dismantling the power of the national operator and opening up the domestic market for new and foreign competitors. Competition in traditional domestic markets and technological developments put international tariffs and profits under pressure and increased the costs of new investments enormously.

European public telecom operators, such as Deutsche Telekom and France Telecom, responded to these fundamental changes by applying a wide range of strategies to strengthen their position in the domestic markets and to expand their operations in new markets (Graack 1996; Noam and Kramer 1994). They became private corporations with stock-market listings and national governments as majority shareholders. Strategic alliances were formed among the telecom operators to strengthen their position in the international telecom market. To compensate for the loss of income in traditional domestic markets many operators introduced new products and services, and searched for new international markets (van Marrewijk 1999, p. 86). Furthermore, internal reorganizations were carried out to transform the former government department into a provider of commercial services.

In order to cope with the consequences of restructuring the international telecom market KPN developed new strategic plans. KPN was the national telecom department in the Netherlands with 33 000 employees, a turnover of EUR 6.5 billion and profits of EUR 1 billion in 1990 (Annual Report 1994). In 1989, the top and middle management of KPN had to gear up for the intended restructuring. The new senior management introduced a new style of management which stimulated responsibility and entrepreneurial skills in the top of the organization. This was necessary because more than 100 years of monopoly had made a deep impact on the organization. The former government department had plenty of experience of government policy but hardly any experience of commercial markets.

The KPN management opted for an ambitious strategy. After being dominated by the Dutch government for so many years the former public-sector organization saw opportunities to become a commercially-driven international telecom service provider. It is for this purpose that a growth strategy was formulated with three themes: (1) build excellent customer relations, (2) ensure access to efficient telecom networks, and (3) expand to new and upcoming markets in West, Central and Eastern Europe, and South-East Asia (*Bedrijfsvisie* PTT Telecom 1995). KPN wanted to become a world-class telecom operator.

After this preparatory period a new organizational restructuring programme, named *Perspectief* (Perspective), was introduced with the aim of redesigning all the processes in the organization. Over the years a new organization was built up with distribution channels, such as Primafoon shops, and customer-care processes. Former technicians were retrained to become sales employees. To satisfy customers and to bridge the wide gap between the customers and the organization, 32 telecom regions, concentrated in 13 districts, were created. With two or three hundred employees each region offered the full package of products and services to a quarter of a million customers. The employees responded positively to this reorganization. As a result, the organization and the customers got to know each other better. To meet the specific demands of the customers the telecom region was separated into three process units: large customers, customer market, and service. To guarantee the quality of the reorganization a total quality management programme was introduced, giving KPN the instruments to structurally measure quality of service and customer satisfaction.

The *Perspectief* programme changed the Fordistic organization of the production. Traditionally, the production process was organized in specialized divisions that were coordinated at central level. Mintzberg (in Steijn 2001, p. 23) calls this Fordistic organization structure, which institutionalizes the social and technical division of labour, a 'machine bureaucracy'. KPN could be characterized as a machine bureaucracy preoccupied with rules and procedures (van Marrewijk 1999, p. 136).

Fordistic organizations are typified by a high degree of internal complexity to coordinate the management and the executives tasks (Steijn 2001, p. 27). They are also characterized by industrial mass production, productivity growth, lifetime employment, collective agreement, planning and control (Trommel 1999, p. 42). *Perspectief* brought the technical division of labour to an end; the production of services and products was no longer centralized but delegated to 32 telecom regions.

The decentralization cleared the way for Operation *Kwartslag* (Quarter Turn) which aimed to trim 25 per cent of the costs of headquarters in 1994. Employees were coached to apply for new positions in the regions or find work outside the organization. The objective of 25 per cent was never fulfilled.

In the mean time, the international ambition of KPN was realized. New offices were opened at high speed in London, New York, Brussels, Tokyo, Milan, Budapest, Frankfurt, Kiev, Jakarta and Prague. A strategic alliance, Unisource, was created together with Swedish Telia and Swiss Telecom to serve the multinational clients in Europe. The Spanish provider, Telefónica, joined this interesting concept and, later on, the American AT&T became the American partner of Unisource (van Marrewijk 1999). The original objective of this alliance was to create one new telecom operator. The identities of the parent companies would merge into a new Unisource identity. In order to prepare for this amalgamation KPN took minority shares in the Czech Republic, together with Unisource partner Swiss Telecom, and in Ireland with Swedish Telia. As a result, KPN became more and more networked into international partner organizations.

Looking back over the period from 1989 to 1995, Verwaaijen, one of the new top managers, said in a newspaper interview (*NRC Handelsblad* 15 June 2000) that he greatly enjoyed this time. Dik, the chief executive officer (CEO), used the metaphor of a mammoth tanker that has slowly but steadily changed its course (Dik 1998, p. 121). By the mid-1990s competition was in full swing in the Netherlands. The organizational costs were still very high compared to those of the competitors, which used simple and cheap processes. Nobody in KPN exactly knew the cost prices of the products and services and which products and services created profits or losses. Accordingly, the organization needed to change fundamentally again. To pre-empt resistance to another reorganization programme, employees were invited to discuss the future of KPN in the *Telecom Overmorgen* (Telecom the Day after Tomorrow) project. This opportunity was positively received by the employees and resulted in new ideas.

Operation *Accent* (Emphasis) in 1997 transferred the technological future of the telecom market into operational demands for the organization. On the basis of the outcome of the discussions in *Telecom Overmorgen* and on *Accent* the *KPN Nieuw* project (KPN New) was formulated. The 13 districts were liquidated in order to gain insight into the costs of products and services, to simplify the processes, to streamline the organization and to reduce operational costs. Control of sales and operations was again centralized to realize greater decisiveness than in the 13 little 'kingdoms'. Business units were therefore set up with responsibility for profit and loss. These business units were responsible for the production of services and products while the sales unit was responsible for sales to the customer. The need to adapt faster to new market trends forced KPN to organize production in customer-focused chain dependencies. KPN Telecom was gradually developing into a multinational with four subsidiaries: KPN Mobile, KPN Data/IP, KPN Fixed Net and KPN ICM (Internet, Call Centres and Media). The *KPN Nieuw* programme introduced the characteristics of a post-Fordistic organization, that is, chain dependency between the organizations (Steijn 2001, p. 28).

However, the introduction of *KPN Nieuw* and the liquidation of the districts went too fast. Employees who used to work in well-organized regions could not find each other in the new organization. This reduced the quality of the operations and the loyalty of employees. The liquidation of the districts forced the employees to relocate along with their work to other places. However, many of them applied for jobs in other companies (KPN *Annual Report 1999*, p. 14). Consequently, customer satisfaction dropped sharply. The *Prisma* (Prism) programme was then started (*KPN NU*, 24 June 2000, no. 46) to improve operational quality and customer satisfaction

As a result of the change programmes, *Accent*, *KPN Nieuw* and *Prisma*, a large part of the new KPN services and products were developed, introduced, and produced together with international partner organizations. Many examples could be quoted to illustrate the change in primary process. For instance, telecom services for multinationals were produced by Unisource, an alliance of KPN, Swedish Telia, Swiss Telecom and Telefónica. Employees of the different parent companies worked together in Unisource to merge the international telephone networks of the four organizations. Another example is the new strategic alliance KPN-Qwest with the American network operator Qwest. This alliance was started so that KPN and Qwest employees could work together to build new data networks and sell capacity for data transport to customers. Among the many other examples is the strategic alliance *Xantic* for the satellite business in cooperation with the Australian Telstra and the introduction of i-mode with the Japanese DoCoMo. These are cases of what Castells (2000, p. 207) calls the 'producer and the customer inter-firm networks'. The production process of KPN products and services was no longer organized within the organization itself, but within an inter-organizational network (see Figure 10.1).

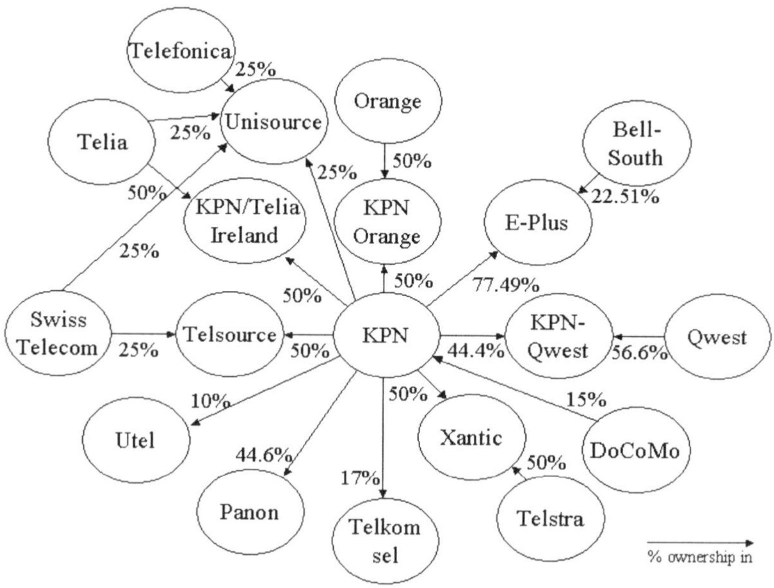

Figure 10.1 Strategic alliances of KPN

Although the company was part of a network with large operators in different market segments, the search continued for a strong merger partner. The merger negotiations with former Unisource partner Telefónica and with Belgacom both failed. KPN's financial partner, Hutchinson, caused the large debt that forced KPN to merge. Together with Hutchinson, KPN acquired a licence for a Universal Mobile Telecommunications Service (UMTS) frequency in Germany for EUR 8.4 billion in October 2000. An escape clause in the contract allowed the Hong Kong-based Hutchinson to withdraw the day after, leaving KPN with an enormous debt. This debt could not be paid off with the stock-market launch of KPN Mobile as the financial climate at that time was unfavourable. Despite good profits, the organization came under heavy criticism and the stock-market prices declined sharply.

In response to these developments the Vision programme was started in autumn 2000. The aim of this programme was to transform KPN into a group of subsidiaries under the supervision of a scaled-down headquarters (KPN *Annual Report 2000*, p. 22). The subsidiaries gained large independence through a clear separation of production and distribution channels. At the same time, a new management model based on uniform and simple production-chain management instead of departments was introduced (*KPN NU*, 11 November 2000, no. 52, p. 4). Six thousand jobs in non-core activities had to be shed. This brings us back to the start of this chapter where I described the consequences of *Vision* for my own personal situation.

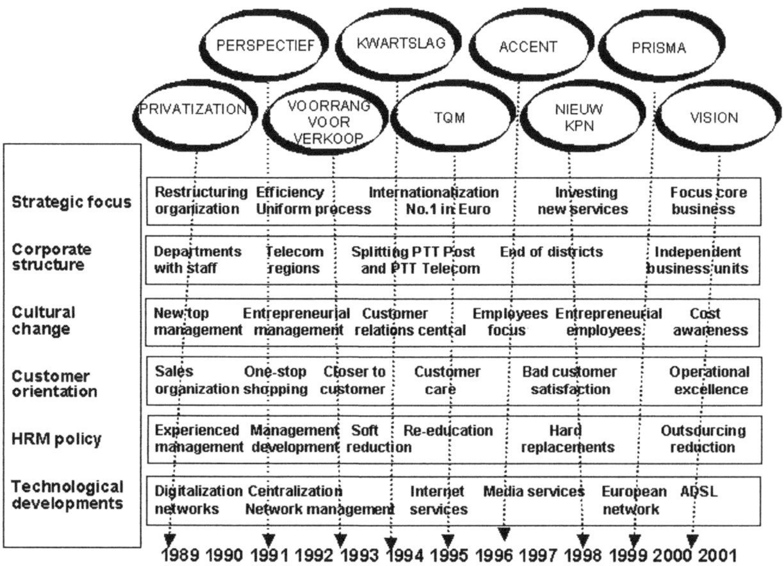

Figure 10.2 The restructuring of KPN in the period of 1989 to 2001

A total of nine different programmes were executed between 1989 and 2001 to change the traditional bureaucracy of KPN (see Figure 10.2). These programmes resulted in several developments in the organization of production. First, KPN was restructured from a machine bureaucracy to an organization with new production concepts, including production chains, total quality management and the flexibilization of labour. The division of labour was minimized with the introduction of telecom regions. Later on, the production chain was redesigned into independent business units with a clear division between production and distribution channels. The second trend was the growth of dependencies on other organizations in complex inter-organizational networks. For the production of services the business units were interdependent and related to other (international) companies. Technology, infrastructure, finance and knowledge were among the contributions needed from the other companies in the network. The third development was the internationalization of KPN's business activities to encompass new international markets in association with other telecom operators. The organization tried, with varying success, to become a multinational corporation. These three developments slowly changed KPN from a bureaucracy into an organization with the characteristics of a network company as defined by Castells (2000, p. 188).

The exploration of the fundamental restructuring of KPN shows that the organizational context had changed dramatically for its employees. How did KPN employees define their identity in this new organizational context?

THE CONSTRUCTION OF IDENTITY AT DIVISIONAL LEVEL

The restructuring of the telecom market changed the context of KPN. On the basis of the conditions for identification as discussed by Pratt (2001), inter-organizational competition with other international operators should have stimulated employee identification with KPN. The employees did indeed describe the organization and its competitors by using the metaphor of a fleet ranged around a mammoth tanker sailing more or less in the same direction and under attack from submarines (van Marrewijk 1999, p. 106). But the first condition of identification, corporate distinctiveness, had become weaker. The development of KPN into an intra-networked organization made it increasingly difficult for employees to perceive KPN as one integrated company. They also used the metaphor of a brick wall that is slowly tumbling down (van Marrewijk 1999, p. 107).

The employees were being confronted with permanent changes in organizational structures, increasing complexity, uncertainties and frustrations. As organizational structures changed regularly, individuals had to become flexible. For example, they had to move abroad with their families, change their home or office, commute long distances, get used to new bosses and apply for new jobs. The employees of the International Consultancy Department had to find new employment:

> It is clear that the current structure of IC is coming to an end; it is a period that has nearly gone. All IC employees will have their own thoughts on this, and follow their own paths. It would be wonderful if a substantial group of IC employees could join the employee or the management buy-out. Others will choose different roads, alone or in a group. What is most important is that you make your decision on the basis of your own values. This difficult period can serve as a springboard for new steps that would not otherwise have been taken.
>
> (Human Resource manager in *Weekly*, a KPN electronic magazine)

The statement of the Human Resource manager that 'What is most important is that you make your decision on the basis of your own values' implies that employees were excluded from the KPN organization. These fired employees could no longer identify with the organizational values, because the nested identities discussed by Ashforth and Johnson (2001) were no longer relevant.

Traditionally, KPN employees identified with the colour green, the company colour. In corporate publications the expression 'I'm sprayed green' was used to express individual loyalty to the organization. In the corporate history, employee loyalty was always highly rewarded (van Marrewijk 1999, p. 136). Now that the organization had been restructured and so many people had been laid off, KPN employees no longer perceived loyalty as important. Steijn (2001, p. 52) calls this the change from the 'old' to a 'new' psychological contract between employees and the organization. In the old psychological contract, loyalty, discipline and obedience were given in exchange for lifetime employment, job-related pay and aged-based career planning. In the new psychological contract, maximum involvement, mobility, responsibility and maximum commitment are asked in exchange for performance-related pay, challenging projects and management development. In the new psychological contract loyalty is not rewarded.

KPN employees were increasingly identifying with innovatively orientated business aggregates. As loyalty to the corporate organization was no longer rewarded, they were looking for challenging projects and career opportunities. KPN employees seemed to identify with their business units and products such as mobile, Internet, fixed net, and data network, but this did not promote internal cooperation within the KPN network. In a study of the internal cooperation of the KPN Mobile and KPN Fixed Net business units, Propitius (2000) found that a difficult relationship existed between the 'arrogant' employees of KPN Mobile and the 'dusty' employees of KPN Fixed Net. Because the mobile telephony department was very successful, its employees labelled themselves as 'young', 'hot', 'sexy' and 'dynamic' (Propitius 2000). The perceived differences between KPN Mobile and KPN Fixed were salient. The KPN Mobile employees perceived the 'old and dusty' employees as working in the traditional telecom business of fixed networks and suchlike. This is how KPN Mobile employees distinguished themselves from other business units. Furthermore, KPN Mobile was located in a separate building with primarily young employees. According to Parker (2000), spatial distance and the generation gap can be used to distinguish employees within an organization. The three conditions for organizational identification named by Pratt (2001, p. 17) were set in the cooperation of KPN Mobile and Fixed Net.

Cultural identity can lead to power struggles (Cohen 1994, p. 119). The external categorization of a group by other more dominant groups can result in the acceptance and affirmation of a group identity in which the dominant group has more power. If this group identity is rejected, it results in a power struggle (Jenkins 1997, p. 70). The Executive Board at KPN understood the need to decentralize operations into independent business units, but it was also struggling to realize the shared identity of these different units and prevent any destructive power struggle. To stimulate shared identity, KPN launched its 'Wij zijn KPN' campaign (We are KPN) in autumn 2000.

As already mentioned Ashforth and Johnson (2001, p. 33) state that lower-order identities tend to be more salient, while organizations invest considerable energy in boosting the awareness of higher-order identities. Before the restructuring, KPN employees constructed their identity primarily at regional or district level, which is also a lower-order identity (van Marrewijk 1999, p. 105). So, the social identity processes in the business units of the KPN intra-organizational network were not new. It could be argued that these business units were the clearest and most stable parts of the network. Were the changes more radical in the parts of the organization where the intra-organizational network was more ambiguous? This question will be addressed by exploring the identity processes in the Unisource strategic alliance.

THE CASE OF UNISOURCE: A VAGUE, DIFFICULT AND UNSTABLE CONTEXT

The Unisource strategic alliance, in which each of the parent companies, Telia, Swiss Telecom, KPN and Telefónica, had a 25 per cent share, is an example of an ambiguous part of the inter-organizational network of KPN. The four parents mobilized their human, financial and technological resources to develop international services for multinational clients in Europe. In their daily activities, the Unisource employees had to work together with colleagues recruited from different parent companies. The inter-organizational differences between the parents were experienced at national as well as at corporate level (van Marrewijk 1999). This made it difficult for Unisource employees to develop a sense of belonging to the organization on basis of shared results. An example:

> When I was in Sarajevo for business the satellite connection between Sarajevo and Madrid broke down. This connection was the only reliable means of communication between the United Nations Developing Programme and the United Nations and EC headquarters in Geneva and Brussels. Late that night, I phoned the Spanish network management centre in Madrid to report the breakdown. At that time no English-speaking members of staff were in the network centre, so no-one realized that there was an emergency and that the line had been broken. I called KPN in the Netherlands and they repaired the satellite connection. I then contacted Madrid via a ground-station in Brussels. 'I am calling via the satellite in Brussels from Sarajevo for a test of the line' I said to the Spanish centre because the line had to be tested from the centre in Madrid. 'No that is not possible, the line with Sarajevo is down' replied the Spanish centre in broken English. 'Yes, but I am calling from Sarajevo, via Brussels' I tried again. It took some time before we understood each other. This caused a day's delay in the repair of the line.
>
> (Dutch Unisource employee, ex-KPN)

In this situation the Unisource employee felt a greater sense of belonging to his former KPN colleagues than to his new Unisource colleagues. In other situations, Unisource employees distinguished themselves by claiming that they were based in Hoofddorp, at some distance from the Dutch KPN. Furthermore, they had a self-image of being young, international, commercially orientated and dynamic while they perceived the parent companies as political, national and bureaucratic. These forms of separation reflect aspects of the spatial/functional, generational and occupational/professional division discussed by Parker (2000).

To promote understanding of cross-cultural relationships and to create a common cultural platform the joint Management Development Programme (JMDP) was set up at Unisource. Managers of the four parents and Unisource managers joined this programme, working in teams at one of the headquarters of the parent companies. A virtual intra-network supported the face-to-face contacts and the personal networks that were built up during the team meetings. Employees were curious about their colleagues from the parent companies and positively identified with the status associated with personal international networks, travelling and international business.

> Finally, I have also been to Madrid for the programme. All of my colleagues have already travelled abroad, except for me. We were met at the airport by the Spanish delegation with black limousines and were taken to an excellent restaurant in Madrid. I had taken my wife along because the event took place in the weekend.
>
> (Interview with KPN manager)

People enjoyed the programme that afforded them the opportunity to travel through Europe, to learn about the parent companies and to build international personal networks. The managers of the parent companies strongly identified with the international character of the Unisource alliance and negatively rated the national character of the parent companies. Pratt (2001) calls this 'disidentification', which is based on identifying with what a group is not. Disidentification is not a break with identification, but rather identification with values and beliefs that are antithetical to those of the group (Pratt 2001, p. 20). This concept created space for conflicting or ambivalent identifications in Unisource in which individuals were both attracted and repulsed by their organizations.

More ambivalent identification could be observed in the Joint Management Development Programme. The programme increased the knowledge on cross-cultural cooperation and in the course of the programme the perception decreased that values were shared (van Marrewijk 1999, p. 174). The confrontation with colleagues from other cultural backgrounds evidently stimulated the employees to reflect on their own cultural identity. The employees of parent companies adhered to their cultural identity and thus their readiness to change habits and learn effectively decreased. The old personal networks in the parent companies were preferred to the new Unisource networks.

> It is quite difficult here; it's like walking in a minefield. You can walk on one side but not on the other. You can be very direct but … you never know when you might be speaking too openly with someone. You don't know where the conversation is passed on or to whom. It is of course the same in any organization but here some of the people have known each other for a long time. You never know exactly how these networks fit together …
>
> (Interview with Unisource employee)

The preference of informal to formal networks can be understood by applying the concept of 'informal cross-cutting identities' (Ashforth and Johnson, 2001). These identities differ from formal nested identities, which are related to the formal organizational structure. Informal cross-cutting identities include the personal networks with former KPN colleagues. A case in point is the 'Voorlindenaren'. These are groups of new KPN employees who have been in the same management development trajectory (van Marrewijk 1999, p. 142). As a consequence of these cross-cutting identities, the connections between Unisource and the parent companies remained strong, leaving the employees space to use different identities in their daily work:

> Formally, I don't have a formal relation with KPN beyond the fact that it is the network supplier, but informally I have my own networks in the organization. I think it's bad, but it's all about who you know in the organization. The informal circuit is much stronger than the formal circuit.
>
> (Unisource employee)

The growth of Unisource was rather chaotic and the institutional relations between the parent companies were unstable. Telefónica, for example, joined Unisource in 1996, but left again in 1998. This departure came as a disappointment to the Unisource employees who had been involved in the cooperation with the Spanish operator. For many, it was a frustrating experience. For three years the Unisource employees had invested large amounts of energy in establishing personal networks, creating new commercial opportunities, changing positions and some had even moved to Spain with their families.

> My colleague moved to Madrid with his family only three months ago. He learned Spanish, rented a house, found schools for his children and was prepared to work over there for a couple of years. Now he can come back and try to find his way again in Unisource.
>
> (Unisource employee)

Frustration slowly started to grow among the employees because of the many changes to the strategy, the unstable character of the partnerships, the difficult corporate politics, the informal personal networks and the chaotic organization. Many of the Unisource employees did not clearly understand how the different parts of the Unisource group contributed to success. Employees started to get confused about the vague strategy of the alliance and the parent companies. One Unisource employee said, 'Sometimes we are just like a bunch of hamsters in a treadmill, running around but getting nowhere'. Employees therefore wanted more information in order to be able to reflect upon their membership of the organization. According to a survey conducted at the end of January 1997, 51 per cent of the 637 employees located at sites in the Netherlands, the UK, Sweden, Belgium and Germany asked for more corporate information from the management on the company course, strategy and business planning, and for more interdepartmental and cross-departmental information. Furthermore, they called for an improvement in the general management style and a clear understanding of how employees fitted in with the overall objectives of the organization (Unisource 1997).

This Unisource case shows that the construction of identity is much more ambiguous than in the KPN case discussed earlier in this chapter. According to Bartel and Dutton (2001), ambiguity occurs when the employee is unclear about individual membership status. Ambiguity existed in the tension between the Unisource and the partner membership. Feelings of belonging to Unisource were not triggered by sharing the results of events between the employee and the organization, because there were differences in the corporate and national culture. Furthermore, Unisource employees found it difficult to evaluate their organizational relationship and develop a sense of belonging due to organizational chaos and lack of information.

CONCLUSIONS

This chapter has focused on the restructuring of KPN and the way its employees constructed their identity. The examination of the restructuring process in the period between 1989 and 2001 shows that the change programmes *Privatization*, *Perspectief*, *Kwartslag*, *Telecom Overmorgen*, *Accent*, *KPN Nieuw*, *Prisma* and *Vision* were executed to restructure the organization. As a result, the organization of production changed from a traditional bureaucracy, characterized by a large division of labour and a complex coordination of tasks, to an organization with features of a network organization of the type discussed by Castells (2000). The production of telecom services was organized into independent business units which were linked in production chains. To produce these services these business units used the technological, human and financial resources of international partners. Production was therefore no longer concentrated in the Netherlands but organized in an international network of strategic partners. Not only production, but also the consumer markets of KPN products and services became international. E-plus for example, the second mobile operator in Germany, is owned by KPN thus giving the organization features of a multinational. The new KPN organization therefore, can be characterized as an international intra-organizational network.

The intra-organizational network of KPN made it more difficult for employees to construct an organizational identity than in the former bureaucracy. The many change programmes, the flexibilization of labour, the massive lay-off, the outsourcing of non-core activities and the financial debts did not provide KPN employees with a positive self-definition of who they were. Identification with KPN corporate goals and values had become much less salient. As the employees worked in business units in a fragmented intra-organizational network they increasingly constructed their social identity at divisional level. This confirms the findings of Ashforth and Johnson (2001) that lower-order identities are more salient than higher-order identities. The importance of the division identity for individuals however, was not new in KPN. Traditionally, KPN employees constructed their identity at district and work group level. To most of the employees, therefore, their organizational memberships in the intra-organizational network were not ambiguous. In the business units the individual membership status of employees was clear.

The case of the Unisource strategic alliance shows that in some parts of the intra-organizational network, the construction of identity was much more ambiguous. The organizational context of the Unisource alliance was somewhat vague, problematic and unstable. Due to cultural differences, obscure corporate strategies, minimal communication, and unstable corporate relations, the construction of identity was difficult for Unisource employees. The sense of belonging to the parent companies created space for ambiguous social identification and situational construction of identity as discussed by Koot (1994).

The concepts of nested identities and cross-cutting identities have been useful for studying the construction of social identity in complex and unstable organisations. As the nested identities were weakly institutionalized in the structure of the organization, the informal identities were important to Unisource employees in their daily work. This is what Ashforth and Johnson (2001) call 'cross-cutting identities' and explains the strong sense of belonging of Unisource employees to the parent companies.

REFERENCES

Ashforth, B. and S. Johnson (2001), 'Which hat to wear? The relative salience of multiple identities in organizational contexts', in A. Hogg and D. Terry (eds), *Social Identity Processes in Organizational Contexts*, Philadelphia: Psychology Press, pp. 31–49.

Bartel, C. and J. Dutton (2001), 'Ambiguous organizational memberships: constructing organizational identities in interactions with others', in A. Hogg and D. Terry (eds), *Social Identity Processes in Organizational Contexts*, Philadelphia: Psychology Press, pp. 115–30.

Castells, M. (1997), *The Power of Identity*, Oxford: Blackwell.

Castells, M. (2000), *The Rise of the Network Society: Second Edition*, Oxford: Blackwell.

Cohen, A. (1994), *Selfconsciousness: An Alternative Anthropology of Identity*, London: Routledge.

Dik, W. (1998), 'Cultuur door de bocht', in C. Quarles van Ufford (ed.), *De Ideologie van de vrije markt*, Amsterdam: Vrije Universiteit Amsterdam Uitgeverij.

Friedman, J. (1994), *Cultural Identity and Global Process*, London: Sage.

Giddens, A. (1990), *The Consequences of Modernity*, Cambridge: Polity Press.

Graack, C. (1996), 'Telecom operators in the European Union; internationalization strategies and network alliances', *Telecommunication Policy*, **20** (5), 341–55.

Hogg, A. and D. Terry (eds) (2001), *Social Identity Processes in Organizational Contexts*, Philadelphia: Psychology Press.

Jenkins, R. (1997), *Rethinking Ethnicity: Arguments and Explorations*, London: Sage.

Koot, W. (1994), 'Ambiguïteiten en wisselende identiteiten: over de dynamiek van organisatiecultuur', *M&O, Tijdschrift voor Organisatiekunde en Sociaal Beleid*, **48** (2), 113–29.

KPN (1999), *Annual Report 1999*, The Hague: KPN.

KPN (2000), *Annual Report 2000*, The Hague: KPN.

KPN (2000), *KPN NU*, 24 June, no. 46, The Hague: KPN

KPN (2000), *KPN NU*, 11 November, no. 52, The Hague: KPN

Marrewijk, A. H. van (1997), 'The internationalization of Dutch PTT Telecom from a cultural perspective', *Telematics and Informatics*, **14** (4), 365–81.

Marrewijk, A. H. van (1999), *Internationalisation, Cooperation and Ethnicity in the Telecom Sector; An Ethnographic Study of the Cross-cultural Cooperation of PTT Telecom in Unisource, the Netherlands Antilles and Indonesia*, Delft: Eburon.

Noam, E. M. and R. A. Kramer (1994), 'Telecommunications strategies in the developed world: a hundred flowers blooming or old wine in new bottles?', in C. Steinfield, J. M. Bauer and L. Caby (eds), *Telecommunications in Transition: Policies, Services and Technologies in the European Community*, London: Sage, pp. 272–86.

Parker, M. (2000), *Organizational Culture and Identity: Unity and Division at Work*, London: Sage.

Pratt, M. (2001), 'Social identity dynamics in modern organizations: an organizational psychology/organizational behavior perspective', in A. Hogg and D. Terry (eds), *Social Identity Processes in Organizational Contexts*, Philadelphia: Psychology Press, pp. 13–31.

Propitius, J. (2000), 'Kiezen of delen; Een onderzoek naar de effecten van intergroepsprocessen op de samenwerking binnen KPN', unpublished master's thesis, Vrije Universiteit Amsterdam.

PTT Telecom, *Annual Report 1994*, Den Haag: PTT Telecom.

PTT Telecom, *Bedrijfsvisie PTT Telecom 1995*, Den Haag: PTT Telecom.

Steijn, B. (2001), *Werken in de informatiemaatschappij*, Assen: van Gorcum.

Tajfel, H. (1981), *Human Groups and Social Categories*, London: Cambridge University Press.

Trommel, W. (1999), *ICT en nieuwe arbeidspatronen*, Den Haag: Rathenau-instituut.

Unisource (1997), *Gateway*, 2 April, Issue 2.

11. Identities and Relationships in the Theatre of the Dotcom Industries

Marcel Veenswijk

INTRODUCTION

According to Castells (2000), a technological revolution, centred around information technologies, reshapes the material basis for society and organization and forces us towards a new distinctive economic system. Central features of this new economy are its informational, global and networked character. Its consequences are supposed to be a significant compression of work-related activities in time as well as in space, combined with an ongoing flow of informational devices (Shapiro and Varian 1999). Through these dynamic and unpredictable flows and relativities, a new information society becomes characterized by the explosive pluralization of subjectivity, interconnection and artefaction, forming a hyper-real marketplace of semiotic exchange where information becomes the transient medium and message (Jeffcutt and Thomas 1998). Successful organizational initiatives that are closely connected to the new institutional contexts – at least until the recent economic 'shakeouts' – are mainly concentrated within the field of ICT, or more specifically the world of the Internet: *The Dotcom* (Kelly 1998). The rationale and core vision of these organizational constructs are, in a way, the opposite of what is 'usual' in the old economy. Negroponte (1995) defines four basic areas in which the differences are exposed: the means, the products, the communications and the markets. At the centre are the means: digital technology opens possibilities for parallel communication, interactions where customers are no longer at the end of the pipeline, but define (in interactive and filtered processes) central product modalities (for example, he/she no longer buys a complete news magazine, but only 'interesting topics'). This form of product assemblage creates new forms of (global and customer-driven) ways of communication and marketing. It suggests a world in which the old rules of 'doing business' have vanished or have at least been fundamentally altered. To follow a more abstract line of reasoning: many new economy organizational regimes seem to be the result of techno-modernization processes in which individuality, rationality, ambiguity and uncertainty become the predominant features.

This contribution focuses in particular on the complex relationships between the new creative Dotcom organizations (occasionally referred to as new economy organizations in this contribution) and their audiences. The turbulent contexts, which are usually dominated by investors and clients from the old economy, form a necessary aspect for many Dotcoms, but are potentially in stark contrast to the picture of Dotcom companies as independent organizations, driven by intrinsic creativity and authenticity, which emerges from countless publications (see for example Shapiro and Varian 1999, p. 43). The central question in this chapter revolves around the consequences of the changing relationships between new economy organizations and their audiences for the identity processes of players in Dotcom organizations. In particular, which identity processes have emerged in the wake of the shakeouts to the key players of the Dutch Dotcom companies.

In this chapter, identity as a social construction process concerns the complex whole of images and discourses which players (individually and as a group) develop in their own organization or work unit.

The research question suggests a player or lifeworld perspective in which identity is conceptualized as a dynamic entity. Identity is a construct that can be created, accommodated and modified by players both in time and in space, depending on the situational definition of reality (see, among others, Parker 2000).

This perspective can be discerned from a view in which identity is defined in light of the integral organizational system and concerns a cluster of images which are communicated to (potential) customers in order to maximize the profitability of an organization. In this context, Bernstein (1986), among others, uses the term corporate or organizational identity.

In order to answer the research question mentioned above, the discourses around old and new economy will be elucidated in the next section. They serve as a context for the more specific environments and interorganizational domains of Dotcom organizations. The subsequent section discusses the core concepts behind the central question of this chapter, and will stop at the ideal-typical characteristics of new economy organizations and at the same time home in on the concept of (multiple) identity. With regard to the theoretical notions, a set of assumptions will be formulated in connection with the praxis of Dutch Dotcom organizations. The penultimate section will present two cases: X.com and

Y.com, which illustrate the complex relationship and dynamics between the environmental context and the organizational domain. In the final section, the central question will be considered again in the light of the empirical material.

THE DISCOURSE SURROUNDING OLD AND NEW ECONOMY AS A CONTEXT FOR DOTCOM ORGANIZATIONS

In past years, the discourse surrounding the new economy has arisen in various social fields and with different interests and resources. The discourse surrounding this theme has a direct link with the organizational practice 'to the extent that it is enacted in the micro-contexts of players'. Thus, the discourse exists neither as an objective entity nor as a set of meanings that people carry in their heads, but in concrete action (see Veenswijk 2001). In principle, the debate is focused on the question of whether the new economy concerns a new principle of free market processes as opposed to a debate on norms and values that reflect our present-day postmodern society.

If we take Castells's view, then the new economy, with its corresponding organizational arrangements, is a phenomenon that has arisen not just in the last few years, but in the last quarter of the twentieth century, because, in his opinion, 'the information technology revolution provided the indispensable, material basis for its creation. ... It is the historical base linkage between the knowledge-information base of the economy, its global reach, its network-based organizational form, and the information technology revolution that has given birth to a new, distinctive economic system' (Castells 2000, p. 77). Although Castells puts the emphasis on the interdependence of globalization, the formation of networks, and the exponential growth of information technology, he stresses above all the last aspect. The developments in the field of information technology in recent years have, in his view, led to an increase in productivity that under no circumstances could be matched in the previous epoch. Vuijsje (2001) attributes the fact that the concept of the new economy is a matter not only of a general productivity increase but also of the emergence of a technological social subdomain to the enormous interest generated by investors for so-called new economy funds that involve organizations active in the area of ICT and telecommunications.

The idea of the new economy as 'successor' and 'alternative' to the classic forms of management has been alluded to previously by Kelly (1998) and Dickson (1990). Dickson suggested as early as 1984 that a new kind of free market orientation had arisen in the new techno-economic era under the influence of ICT developments. Dickson defines the new economy movement in terms of *a collective corporate techno-culture for the postmodern capitalism.* His work raises several arguments for the way new economy can be analysed as a representative of the new technoculture. For instance, he points out that such an umbrella reference framework is attractive to industrial leaders, whether it is now called new economy or technological turnaround. For example, it provides a common management language, which, in contrast to the Taylorian 'command and rule' communication, can be used for discussions and negotiations at all levels of the business. At the same time, it does not call into question the collective strategy of senior management. This technological collective culture enables the decentralization not only of the daily decision-making process, but also of strategic planning, by means of a management language that gives both an element of priority (see Dickson 1990, p. 107). In addition, he points out that a broad range of control technologies can be identified under the influence of a techno-culture related to the new economy. After all, the software used for automated production processes provides parameters for monitoring industrial processes and measuring productivity. In this context, Zuboff (1988) refers to the *panoptical power* of information technology.

Besides the productivity aspect and the emergence of the techno-culture, another dimension to the new economy is exposed by Pine and Gilmore (2000) who draw attention to the adventure and the experience that accompany new economy services. They argue that in this age entertainment organized on a large scale (that is, television, films and the Internet) is reaching an ever wider audience. This leads to a new form of 'fun' consumerism that is predominantly rooted in the ICT sector. In the authors' view, this has a direct influence on the action models in organizations: *direct the employees to act.* By this, they are referring not metaphorically to Goffman's (1982) theatre, but literally: it is not a metaphor but a model for action. Whenever employees are in front of customers, they are acting, whether they know it or not, whether they do it well or not, they are nonetheless acting (Pine and Gilmore 2000).

It almost goes without saying that 'acting out' customer relations is done most easily through specially construed systems, for example, Internet websites. If we take Henshall's view, then the new economy goes one step further. In particular, the business models as such are strongly changed as a result of the internet influence. Or as he puts it: 'The internct is changing business models and empowering consumers to create new communities that combine the power to aggregate rich sources of individually personalized data in real time activities. Large-scale data aggregators are emerging to navigate and mediate info-markets' (Henshall 2001). The ways in which value orientations of these business models continue to work in the profile of Dotcom organizations will be elaborated below.

NEW ECONOMY AND DOTCOM ORGANIZATIONS: THREE BASIC CHARACTERISTICS

Despite the shakeouts at the start of the new millenium and the huge decline of the new economy linked ICT sector, the removal of the concept of the Dotcom organization has become inconceivable in present-day business life. In particular, the (alleged postmodern) business concepts developed from the Dotcom sector and their inherent virtuality appear in the mean time to have taken on increasingly fixed forms and show a strong relationship with knowledge-intensive ad hoc companies, such as advertising agencies and research and development departments (see the virtual organization). Three aspects connected with the basic character of the Dotcom organization will be discussed below, namely (1) the partial lateral and virtual nature of the interorganizational network domain of the Dotcom organization, (2) the multiple inclusive player identities, and (3) the 'here and now' connection and involvement of shop floor members.

Supply-Orientated Virtual Environment

The inter-organizational contexts in which Dotcom companies function are partially visible and are for the most part of a lateral and virtual nature. Services are now offered on line, where the relationship with customers is always based on (lateral) contact and communication is likewise carried out via e-mail or other 'long distance' instruments (virtual). The sharp growth in the automation of internal and external communications has meant that these Dotcom organizations have potentially become open networks, characterized by the input and output of continuous flows of information and people. The related social activities are no longer tied to a specific location, and contacts have, to a certain extent, become virtual. Efficiency, planning, rationality and control have become the key words for managers in contemporary situations, but at the same time the growth of internationalization and the influx of migrants confront them with different, opposing views about how to carry out projects, such as letting processes take their course and not forcing things too much, assuming an attitude of detachment, taking things calmly and trying to achieve a consensus in decision-making (Veenswijk 2001).

These phenomena in Dotcom organizations cannot be isolated from the cultural processes that are taking place in the broader (post)modern society. There, too, various analysts have pointed out the increase in turbulence and complexity (see Chia 1998; Koot 2003; Martin 2002).

Giddens (1990) refers to three factors which have brought about this complexity: distancing in time and space; the breakup of all kinds of traditional systems; and the growth of reflection on action as a result of the drastic expansion of transnational contacts in the economy, politics and tourism, and the information on other cultural systems offered to us through the media and the Internet.

A consequence of this increasing complexity is that the reality usually presented by Dotcom companies 'out there' as a coherent corporate identity, will in part be considered separate from views, value orientations and corresponding identities that (on the level of the groups and players) are prevalent within the organization. Alvesson (2001) points out, however, that such separate identities also appear in related company types (such as knowledge-intensive organizations). The consequence of this is that, depending on the employees, the organization may show to a greater or lesser extent a schizoid character (see also Kets de Vries and Miller 1984), or at the very least will experience a powerful tension between front-stage and back-stage reality.

Multiple Inclusive Organizational Identities

Contemporary theorizing encourages us to think in pluralistic and non-essentialist terms; identities are assumed to be multiple and contextual, therefore they must be constructed and secured. Identities are constituted, negotiated, reproduced and threatened in social interaction, in the form of narratives and also in material practices. Within the various constellations, alternative, potentially competing, definitions of reality are able to develop. The result is that the situation determines which definition of reality is used by players when communicating (Alvesson

2001, p. 1105). In this sense, Weick (1979) speaks of multiple inclusion. In patterns of continuous interaction, each player is constantly included in more than one social context. Those other contexts are always in the background. A player can bring definitions of reality that have been developed in a certain configuration into other configurations of which he or she is a part. The construction and reconstruction of realities is influenced by the players' multiple inclusions in configurations.

Identities are, at least partly, developed in the context of power relationships (Alvesson 2001). In particular, this last aspect is essential for Dotcom organizations. Many Dotcoms are organized along the lines of a 'donorship', where several investors interface with the organization and its environment. These donors are certainly crucial in the initial phase of the e-business, and they are a dominant factor in the production of identity, mainly because, from the outset, the number of customers is limited.

As mentioned above, it is of great importance to draw a distinction between identity processes that are linked to groups of players in the organization and the identity components that can be attributed to the organization as a whole: the corporate identity. In the case of Dotcom organizations, this is of particular importance because employees can be linked to the organization through a loosely coupled (virtual) system, resulting in hardly any interactions with colleagues. Therefore, the identification processes at the level of the group are potentially restricted in the sense that individual players constitute their organization-related identity on the basis of general front-stage images of the organization.

A 'Here-and-Now' Relationship and Limited Trust from Shop Floor Workers

Temporariness is a dominant theme in considering the new economy and Dotcoms. Shapiro and Varian (1999) point out that the reasoning that invariably underlies the selling off of services, appears likewise to prevail in the business models. Employees, like virtual services, can be bought and disposed of (Norton 2001). Several authors have confirmed a similar vision from the employees' perspective. The process of individualization results in employees preferring in the first instance all forms of temporary relationships to unilateral traditional forms of commitment (see also Vuijsje 2001).

Giddens links the tendency towards individualization with a strongly declining trust in long-term commitments, whether these happen to be economic, organizational or social in nature. He claims that insecurity has arisen because the process of individualization has led to a decline in the importance of all kinds of traditions, kinship networks, local communities, and religious forms of association, resulting in turn in a less normative creation of order and, thus, a decrease in the unquestioned nature and predictability of behavior (see also Koot and Veenswijk 1999).

Although Giddens's statements cover a large spectrum, it is obvious that in the context of Dotcom organizations, the question of temporariness and trust plays a specific role, in view also of the restricted nature in which these organizations are generally institutionalized.

From Conception to Presupposition

On account of the (ideal-typical) basic characteristics of new economy organizations that are being discussed here, the following assumptions can be derived:

1. From Giddens's views concerning the distancing in time and space and the breakup of all kinds of traditional systems, as mentioned under the first basic characteristic (concerning the virtual character of Dotcoms), it can be construed that virtualization of organizational domains may lead to a further pluralization of subjectivity, with a strong focus on interconnection and artefaction, and consequently the forming of a hyper-real marketplace of semiotic exchange.
2. The multiple relationship ties that are inherent in the type and functioning of Dotcom organizations, such as those emerging from basic characteristic number two (concerning the multiple inclusive identities), potentially encourage Dotcoms to operate increasingly as open networks, where members of the organization create and modify their group identities in relation to focus points inside or outside the organizational domain.
3. The assumption that can be derived from the third basic characteristic is that Dotcoms in their existence portray themselves potentially as temporary meeting places of 'passers-by' who find each other from their own rationality and where values such as togetherness and bonding are weakened. This favours a practice in which individuality, rationality, ambiguity, and uncertainty have become predominant.

In view of the limited knowledge that is currently available on processes of virtualization in relation to processes of group identity in organizations, named assumptions have a strongly heuristic character. After the presentation of the X.com and Y.com cases, they will mainly be used as a guiding principle for evaluating the empirical material from this contribution.

DOTCOM IN ACTION: THE STORIES OF X.COM AND Y.COM

This section presents two accounts of new Dotcom organizations. The first account focuses on the way the senior management of X.Com deals with the issue of identity at the level of key players in and around the organization. The X.Com management philosophy is closely connected with the idea that organizations are or should be (temporarily) adventures, and management is about guiding and facilitating the participants (co-workers and customers). In this case, there is a significant distance between front and backstage reality.

In the Y.com story, the relationship between front-office workers and the (long-distance) client systems is evaluated more specifically. This case concentrates on the way communication is organized in terms of identity, quality and image. The account shows that quality definitions from product suppliers strongly influence the identity process of the (intermediary) Dotcom officers. In this context, the logic of network partners can become an iron cage for the Dotcom firm.

X.com: Rise and Decline of an Accelerator

The study at X.com has been under way for some considerable time. Until now, we have spoken mainly with representatives of the founding generation: enthusiastic (ex-)students, who are spellbound by the hacking and the attractive products that result. The most important discovery is the accelerator where information can be looked up on the web and classified with incredible speed. Although this product has not as yet yielded any customers, in the mean time there is a staff of about 45 working for the organization. Each month, the salaries are paid by the investor, who is convinced (although ever less so) that sooner or later the customers will automatically come. One morning, we meet Dave. He is in all respects an outsider. He introduces himself as the new Head of Operations. He is dressed in a smart suit, is in his thirties, has an awkward appearance, and comes from Israel. He is not an (ex-)computer science student, but has made an impressive career with the Israeli army in the Special Forces. Although it is not permitted to speak about the most important investors of the company, it soon becomes clear to us that they are no longer from the Netherlands.

A short overview of the organization

Despite the fact that X.com was started in 1997 from a student dormitory, the company grew in a period of less than three years to a medium-sized organization with 45 employees. The average age is 28 years. The company focuses on developing and delivering knowledge management components that help companies unlock and enrich the information available within the organization, thus making it easier for employees to find and use that information. Its components are designed according to the Component Based Development methodology, and adhere to the Enterprise Java Beans (EJB). With its product, X.com targets companies that need to structure and share large amounts of unstructured textual data, such as banks, publishers and so on. At present, preparations are being made by the management to soften the blows caused by the bankruptcy both for the organization and its staff.

Episode 1: emergence: the accelerator adventure (1998–99)

It just began as a wild plan, a dream. In 1997, four computer science and economics students travelled with a rented car through the USA. During the trip, Silicon Valley fever started to spread rapidly. The idea of setting up their own Dotcom organization gave the group an 'enormous kick'. On their return, the group decided to plunder their bank accounts in order to buy the necessary equipment to make the product that the group had already discussed several times: an accelerator capable of classifying large quantities of information. The accelerator was to be the lottery ticket which, as one of the members of the group put it succinctly, would enable them to win the *jackpot. Hit and run* was the idea. The group wished to make something nice and then expand their horizons afterwards with big bucks. They made the joint decision that each of the four members would be given a suitable task: one would oversee the management of the company, the other the technical side and the others the marketing or the finance.

Shortly after the start of the company, a business plan was drawn up and several presentations were made to potential investors. The investment company, Twinning, found the ideas behind X.com promising, and together with a 'grant' from the company Newconomy, the group had a starting capital of almost 3 million euros at its disposal. In exchange for this, regular progress reports were required and audits on behalf of the investor were authorized. In a very short time, more than ten staff were hired, mostly from the circle of friends of the group. The manager, who in the mean time had been appointed CEO, sought publicity, and before long the organization became known in the Dotcom circuit.

As the number of employees grew, the contours of a corporate culture began to form. The founders were very keen on the organization remaining above all relaxed, and they hammered this home constantly in the interviews. This meant that working hours were not strictly defined, and there was no specific dress code. Instead, an attractive and challenging product had to be made. The Marketing Manager intended it to become a *big climax*. However, there were still clashes. Although at that moment there was still little pressure to actually go out and sell services, the theme 'customer service' was an important bone of contention.

The dominant group had a good deal of confidence in their own abilities and expected that clients 'would value this in the long run'. The Marketing Manager was in favour of active acquisition, although he did not know exactly how to go about it. A year later, the Supervisory Board, in which the most important investors convened, ascertained that there were no concrete customer leads yet. As a condition for a new round of financing, they demanded adjustments to the marketing plan. A member of the Supervisory Board was to be 'actively involved'. This decision led to heightened tensions within the X.com management team. One member of the management team expressed it as follows:

> We had the idea that it was no longer our baby. Grasping back control of the helm would be a great blow. And the blows would rain harder as long as success was missing. I slowly began to feel that I was part of a film, the script of which I no longer had in my hand. This was certainly not in accordance with the picture that I had at the start.

Episode 2: the organization grows up (1999–2000)

In the mean time, the organization was growing. The hype surrounding Dotcoms began slowly to take on astronomical proportions in investor circles. The management team was not insensitive to the sustained mood of jubilation regarding the successes of the new economy. They all calculated one afternoon that the value of their share package in the company had achieved that cherished million. The CEO was wondering whether he could leave the company. The idea was rapidly dismissed by the others. Nonetheless, a new tone was set. The organization increasingly became a means by which the management team could realize their own individual ambitions. The common dream began to wane. The organization had almost 30 staff, and there was a great need for a Human Resources Officer who could take care of the employment contracts.

The Human Resources Officer who was subsequently hired, revealed in an interview that she became acquainted with X.com through being a friend of the CEO. Her picture of the organization was minimal. 'I knew nothing about computers and at that moment was working as an Aids Consultant in a hospital. When I met them, there was a group of men frolicking around in shorts [it was mid-summer] discussing a new knowledge system. That appealed to me.' During her induction period, she realized that in practice staff members were looking for something far more concrete in their employment contracts than had initially seemed to be the case. The desire for security in the field of employment legislation was, according to her, connected with the demands that mortgage lenders made on the staff in granting loans for apartments and other property. She admitted that the organization, in her view, 'did not differ much from the Internet in essence. You siphon off what you need and then you continue surfing'.

In this period, only a small group was aware that the organization merely existed thanks to financial assistance from investment partners.

More and more permanent workstations were added, where care was taken to reserve space for hacking, which, in the view of the programmers themselves, formed not only the core but also the romantic aspect of the work. A programmer formulated it as follows:

> Hacking? By this I mean programming with a group *on the go.* The nicest thing is when you can just make a batch which has not been conceived before. As far as I am concerned, that's . . . Then in the morning you're one of the first up. When the first people roll in, I'm making printouts in a complete daze. Of course, it goes without saying that it should bear fruits for the company.

Contrary to the expectations of some of the staff, work done at X.com was rarely undertaken from private homes. Hacking was seen above all as a group activity. From the very exchange of ideas and emotions (namely, from programmers) a good deal of satisfaction and sense was attained. During the communal (monthly) reunions, the CEO constantly emphasized in different ways the adventurous and meaningful character of the organization. Uncertainty about the future was portrayed as an important asset, a challenge preventing the organization from dozing off.

At the end of the year 2000, an increasing number of reports appeared stating that X.com would have difficulties obtaining another round of financing, as long as no products had been sold. Panic appeared to strike, and a (new) marketing strategy was conceived.

Episode 3: the bubble bursts (2001–02)

Until then, the motto 'live and let live' had been accepted as the dominant leadership concept. Employees were given much freedom to develop their own initiatives, and in this manner to bind themselves to the company. The sustained mood of jubilation in the Dotcom world also included this definition of reality: nice ideas were always rewarded (by investors and customers), and why should this not be the case within X.com itself? In the last phase there was, however, a change in the role and approach of the management. The shakeout that was beginning to take shape as a result of investors pulling out, forced the management to make some internal pragmatic decisions. The main question was for how many weeks the company could still make ends meet without new financing. In this phase, the management and programmers were drifting further and further apart. The management was increasingly seen by the hackers as a messenger of doom constantly invading the creative design process. On the other hand, the management was constantly seeking new schemes to help sales of the products designed earlier. It was a desperate attempt to convince new investors of the feasibility of the organization.

In this phase, both parties became more and more opposed to each other, and many attempts were made by individuals to protect as much as possible their own social environment. This meant for the programmers a further estrangement from everything in the outside world. They withdrew to complicated technical questions and reassumed during communal reunions (dinners and so on) the picture of the ideal creative organization. There were staff changes in the management team, whereby two of the original founders stepped down to give way to an interim manager brought in by an investing partner.

In the end phase of our study we met Dave. Although he would not reveal his background, he did disclose that he had gained experience in an Israeli electronics company, after having made a career in the Israeli army and holding a security function in El-Al. Dave believed that X.com had the potential to develop into a structured organization with clearly defined tasks. Before this could be achieved, the necessary adjustments would have to be made to the management approach and style. Dave also admitted that X.com was just as likely finally not to make it.

Y.com: Clients and Clientele

Only a short time after starting the study at Y.com, an interview was held with a key player in the company. During the conversation, the interviewee was guaranteed confidentiality. The employee conceded during the interview that her work was mainly directed to the building of web applications and that client contact was mainly done via the Internet. Besides her function as a web builder, she spends her evenings further refining and maintaining her own website. I asked her for her web address, because I was curious to know whether the communication (website) at her work differed greatly from her private site. An hour after the end of the interview, I visited her website. This site contained not only a literal report of the interview, including her assessment of how it had gone, but in the mean time also hyperlinks to my university, my teaching group, my curriculum and my publishers had been added.

Short overview

When I went to see Y.com, the company had 68 employees, 60 more than in 1995 when Y.com was just one year old. Y.com presents itself as a full-service office that conceives, builds, hosts and administers advanced online solutions. The designs are built on a multi-system architecture based on J2EE and XML technology. To this end, the organization employs web designers, programmers and technical account managers.

At the time the company comprised three departments. Interactive Concepts deals with business analyses which result in corporate images and in particular related online concepts. The Web Development Department conceives the formation and interaction models of websites. Hosting Facilities arrange for the 'continuous development' of online adaptations. Site security is also found here. The important clients are a Dutch bank, a telecom provider, an oil company and an insurance company. Y.com is located in a historic building in the centre of Amsterdam.

Episode 1: from customer drive to controlled growth (1994–97)

Whereas X.com's focal point was the quality of its unique product, from the beginning Y.com's work took on significance by virtue of the fact that they provided what the customer wanted. Y.com was started in 1994 at the instigation of a large Dutch bank. In this period, the Internet was considered 'something new', and most large organizations wished to participate in some form. There was, however, a general lack of know-how and human resources. As a result of the friendship between the original founder of Y.com and the Communication Manager of the bank, after several preliminary discussions the former was requested to make a presentation to the bank to explain what the Internet could mean for them. After this, several small assignments followed, and after some time a request to build a website. Through contacts with the bank, the number of clients grew immensely in a short period. Six months after the start, eight people were working full time developing and building websites. One of the founders remembered this first period as both a time of great dependence on the client and a period of boundless possibilities. Or, as he put it:

> What we were doing looked really good. And if a large company liked it, why wouldn't other clients? If we really considered our business, we had hit the bull's eye. The only question was how we could show this to the outside world. This was the catch. We knew that it would have to occur through exposure from the bank.

Although the organization attempted as much as possible to remain independent of the bank, it was initially extremely difficult to find similar clients. In a short time, cash-rich competitors appeared, who with their ICT and/or consultancy background, and an extensive media campaign, were able to gain the trust of 'the big players'. Within Y.com, it was soon realized that exponential growth was not part of the package. They opted instead for a controlled growth scenario, in which they would try to expand gradually through the establishment of strategic relationships with possible partners. In 1997, it was decided to enter into a collaborative partnership with two professional advertising professionals. The two became part owners, and, in a short time, the emphasis was put on the marketing of the company.

Internally, a division arose that separated the different aspects of the product according to function. A clear division was created between employees involved in the creative process and programmers and sales representatives. Within the creative group, the prevailing attitude was that a product could only be presented to clients if it met the standards of originality and aesthetics. Many employees – most of them had been to art school – regarded the work as a paid form of applied art that should conform to the highest artistic standards. The Account Manager, who had been with the company since 1998, ensured on the other hand that the demands and desires of the clients were the overriding point of departure. It seemed increasingly difficult to accommodate the different sections of the group, despite the common end product. Or, as a web designer put it:

> At one moment, the whole group decided that no people from the Accounts Department could enter our corridor. It was just forbidden territory. We put up a sign for them: do not enter. However, you shut yourself off, of course, from another part of reality. Money must come in. We simply didn't know how we should deal with it, people worked like crazy. Then management intervened, and after several discussions, relations with the Account Department began to improve.

To strengthen the internal relationship, a strategic vision was devised with the emphasis on working together with external partners. Takeovers and mergers were also taken into consideration.

Episode 2: becoming networked: takeover of a quality player (1998–2001)

In the mean time, Y.com was compelled by three large clients to fine-tune its service package in order to handle specific kinds of requests that regularly came from customers. They were also forced by the client to establish an account forum to ensure that requests from the 'main clients' could be dealt with at all times. Discussions between the management of Y.com and two clients led to an internal *business process redesign*, in which the interactive client concept was to play a central role. Clients also demanded sufficient extra capacity to guarantee customer demand. This, in turn, had all kinds of consequences for the internal company leadership. The Y.com management team was feeling increasingly hemmed in by the peculiar relationships, and they developed three strategies for coping so as to monitor as well as possible their independence vis-à-vis these parties.

First of all, they attempted to stretch their network of relationships as far as possible through entering into a series of virtual association links. Communal workshops with potential partners were organized and they also endeavoured to acquire new assignments through all kinds of virtual interaction, but without success. One of the directors described it as follows:

> Of course we realized only too well that we should be more concentrated in a niche. Together with client contacts, we organized several afternoons to discuss new work forms. It involved a lot of time and money, but brought only

disappointment. Because you do hope that something will come out of it. However, in practice, people let themselves be terribly spoilt for one day and subsequently this becomes the norm. You are then one of the many parties again who may get round to doing something.

Besides partnerships with comparable companies, the management planned to take over Z.com, a company with 15 employees that hardly posed a direct threat to the dominant culture in Y.com, but which had access to a large range of clients. Discussions with the Managing Partner of Z.com resulted in an integration strategy, in which Z.com would be entirely integrated within the Y.com organization. Thirdly, a collaborative partnership with a large ICT company was entered into. Initially, 50 ICT consultants were hired in from the company because of the bank account. The underlying reason was that a major change to the hosted bank could no longer be tackled with the existing structure. The aim of this deal was to be in a position in the long run to profit from clients from the partner's ICT network.

Episode 3: the network contracts

The strategy that was followed had important consequences for the internal culture. From the moment the ICT consultants were placed within Y.com, a strong division arose between account-orientated employees and other staff 'back-stage'. The move of Z.com personnel only served to heighten the them and us ambiance even more.

The designers withdrew into their core activities and met with each other in small groups outside the workplace. In this phase, there was also a split in the management team. Until then they had been operating with a strong consensus model, but now the management team meetings were becoming polarized, and department heads were increasingly representing their own group interests. A member of the management team put it as follows:

We could see it coming, it was already hanging in the air. Members of the management team began to behave more and more like leaders of their own club, the group feeling increasingly disappeared. That also applied to myself. This was also imposed by the departments. The manager simply received a message from his group, which, of course, was not a good thing. When a delicate matter was on the agenda, such as extra resources for a department, this fanned the flames. I felt that I couldn't go on any longer and at a given moment drew my own conclusions.

Instead of a common new identity, the groups crumbled into fragmented subcultures, which hardly struck a chord with the outside world. As a result, the network of relationships was narrowed down rather than expanded, as had been initially intended.

The consequence of this was that once again the dependence on the main clients (on account of the rise in staff numbers) further increased. In view of the increasingly tight financial situation, the Y.com management agreed, after two rounds of negotiations, to the bank taking a substantial financial share in the organization. From that moment onwards, a situation arose in which Y.com positioned itself increasingly towards the interests of the bank. A number of first-wave employees turned against the organization and went to work elsewhere, at the expense of 'the new branch of the bank', which is how Y.com is now referred to.

REFLECTION AND CONCLUSION

At the beginning of this chapter, the question was posed as to what the consequences would be for the changing relationships between new economy organizations and their audiences after the shakeouts of recent times for the identity processes of players in Dotcom organizations.

In the case of X.com, it emerged that the public initially restricted itself to the parties who financially supported the dream of the original founders. In other words, the investors were part of the dream for success and featured as general representatives of the 'there's no end to it' feeling of the new economy. In the terms of Janis (1982), a classic process of *entrapment* emerged: in view of the decision to focus entirely on the accelerator, other stipulations were no longer put on the agenda of the group. A lack of forced correction strengthened the self-referential character of the group initially even more. New employees hardly had a clear overall view of the relationship between the organization and the environment, and they were socialized with the picture of success, which they in turn (necessarily) reproduced in the group. While this success remained elusive, however, and the reserves were exhausted, the central meaning lapsed and the organization became a goalless instrument in the hands of any party that was prepared to inject new life into that story. In the case of Y.com, the situation initially appeared to be just the opposite. The organization was very strongly focused on clients and, attempted as well as possible to associate themselves with the large parties in the market. When the main client took over a substantial part of the shares,

Y.com was threatened with losing its independence. From this standpoint, there were important parallels with the X.com case: the organization was ultimately a plaything in the hands of the dominant party.

What can we now conclude from the cases? The assumptions stated in the third section, can only be partially confirmed in the cases discussed.

The first assumption that virtualized work connections lead to the formation of hyper-real marketplaces of semiotic exchange, and to a pluralization of subjectivity found limited support in the cases. For as far as we see, pluralization in the cases, developments in terms of fragmentation in X.com and Y.com are strongly connected to the *groups* of professionals in the organization, such as the programmers, the account managers and so on. This conclusion is not new. Authors such as Van Maanen and Barly (1985), Martin (1992) and Alvesson (2001) point out that cultural diversification is closely connected with the functionality of groups in organizations. In this sense, the assumptions that are connected to the work of Negroponte (1995) and Shapiro and Varian (1999) underexpose the influence of group processes in new economy dynamics, and may be too much focused on the impact on the individual actors.

As we know from, for example, the work of Parker (2000), the identity construction of players in and around organizations is connected with the higher- and lower-order focus points in the lifeworld of these actors. The cases show that the incentives to connect and identify with specific subgroups in the organization are closely related to the external connections on the level of the organization as a whole. As the diversification and attraction of external donors and customers remained elusive, and ambiguity about success and scope of services increased, players within X.com and Y.com, especially in the creative functions (the web designers and programmers) no longer constructed their own organization (as a key partner for the external groups) as the highest order point of identity. Instead, their own functional subgroup was initiated as the primary focus of identification and created a powerful substitute for what the organization could offer in terms of collective identity. One qualification has to be made: in the front stage of the day-to-day praxis, the linguistic codes that referred to the organization as a whole were still part of the repertoire: subgroups continued to use concepts such as service, quality and authenticity, but these concepts have a specific, (back-stage) meaning, dependent on the context and subgroup in which the concepts are used (for the account manager, authenticity was a selling term, for the web designer a basic criterion for his or her work approach).

In both cases, we can see that players got 'in a fix' when it came to the identification with reference groups.

The second assumption regarding Dotcoms as open networks was also only partly confirmed in the cases. In the case of X.com, in the course of the incubation process an increasing cultural contraction: they withdrew into hacking and gradually lost contact with the outside world. The management attempted with all their power to intertwine the organization into a product-related niche, but each time this failed. Y.com also showed a form of schizophrenia: the designers cut themselves off and re-created once again with each other the story of the creative product, while the account managers identified themselves mainly with the client. A situation arose here that led to internally segmented worlds, where separate discourses prevailed. In any case, the Y.com management apparently realized that they had to position and present themselves in such a way, so as to evoke sufficient initial confidence in their partners.

With respect to the third assumption, which held that Dotcoms were temporary meeting places of passers-by who find each other from their own rationality, just as few empirical references were confirmed in the cases. It is remarkable that staff members in both cases have embarked on surprisingly long-term relationships with the organization. Despite the discourse surrounding the temporary relationship and the notion that you could find a job anywhere, there was little employee turnover (of their own accord). Most of the staff tried to obtain a contract for an indefinite period. In this context, even virtualization did not play an essential role. Indeed, communication was largely carried out by e-mail, but they visited each other in person as much as possible. The offering of the portfolio on line was regarded more as a medium for exposure than as a fundamentally altered way of communicating. Even in contacts with customers, meetings were kept minimally virtual: meetings were held 'face-to-face' in order to gain confidence.

Although the above-mentioned cases barely illustrate the ways in which Dotcom organizations portray themselves against the background of the new economy, the cases show remarkably many similarities with traditional start-ups, such as we know from relevant literature about orgnizations in the old economy (see, for example, Mintzberg 1979). We also find in other sectors the cultural hybridization that arises when front-stage logic is separated from the back stage by pressure from external partners and customers (see, among others, Veenswijk and Stobbe 2001).

What remains of unique Dotcom characteristics, such as the strong longing for renewal, creativity and comparison with paths already trodden, arguably shows many similarities with sentiments from our Western youth culture. It is attributable less to the singularity of the postmodern Dotcom organization because the latter probably never existed in its perfect form!

REFERENCES

Alvesson, M.(2001), 'Social identity in knowledge-intensive companies', *Journal of Management Studies*, **37** (8), 1101–23.

Bernstein, D. (1986), *Company Image and Reality*, Eastbourne: Holt, Rinehart and Winston.

Castells, M. (2000), *The Information Age: The Rise of the Network Society*, vol. 1, Oxford: Blackwell.

Chia, R. (ed.) (1998), *In the Realm of Organization: Essays for Robert Cooper*, London and New York: Routledge.

Dickson, D. (1990), *The Death of the Spirit*, Amsterdam: De Balie.

Giddens, A. (1990) *The Consequences of Modernity*, Cambridge: Polity Press.

Goffman, E. (1982), *Interaction Ritual*, New York: Pantheon Books.

Henshall, S. (2001), 'The consumer manifesto', *First Monday*, (1), 1–16.

Janis, I. L. (1982), *Groupthink*, 2nd revd edn, Boston: Houghton Mifflin.

Jeffcutt, P. and M. Thomas (1998), 'Order, disorder and the unmanageability of boudaries in organized life', in R. Chia (ed.), *In the Realm of Organization: Essays for Robert Cooper*, London: Routledge, pp. 67–87.

Kelly, K. (1998), *New Rules for the New Economy*, New York: Viking Press.

Kets de Vries, M. and D. Miller (1984), *The Neurotic Organization*, San Francisco: Jossey-Bass.

Koot, W. C. J. (2003), 'Hybridisation and disidentification: fatal strategies of modernisation in organizations', in P. Jeffcutt (ed.) *The Foundations of Management Knowledge*, London: Routledge.

Koot, W. C. J. and M. Veenswijk (1999), 'Attachment or detachment, the dilemma of the new manager', conference paper, CMS, Manchester.

Martin, J. (1992), *Cultures in Organizations*, New York: Oxford university Press.

Martin, J. (2002), *Organizational Culture: Mapping the Terrain*, Thousand Oaks, CA: Sage.

Mintzberg, H. (1979), *The Structuring of Organizations*, Englewood Cliffs, US: Prentice-Hall.

Negroponte, N. 1995), *Being Digital*, New York: A. Knopfler.

Norton, R. D. (2001), *Creating the New Economy: The Entrepreneur and the US Resurgence*, Cheltenham, UK and Northampton, US: Edward Elgar.

Parker, M. (2000), *Organizational Culture and Identity*, London, Thousand Oaks and New Delhi: Sage.

Pine, B. J. and J. H. Gilmore (2000), *In Search of the Experience Economy: Work is Theatre and Every Business a Stage*, Harvard: Harvard Business School Press.

Shapiro, C. and H. Varian (1999), *Information Rules – a Strategic Guide to the Network Economy*, Harvard: Harvard Business School Press.

Van Maanen, J. and S. Barly (1985), 'Cultural organizations, fragments of a theory', in P. J. Frost, L. Moore, M. Louis, C. Lundberg and J. Martin (eds), *Organizational Culture*, London: Sage.

Veenswijk, M. (2001), *The Dynamics of Cultural Change in Public Organizations*, Amsterdam and West Lafayette, US: Rozenberg.

Veenswijk, M. and L. Stobbe (2001), *Management van cultuurverandering*, Delft: Eburon.

Vuijsje, F. (2001) 'Overdreven spierballenvertoon, nog een keer: die nieuwe economie', *Intermediair*, **29** (19-7), 25–9.

Weick, K. E. (1979), *The Social Psychology of Organizing*, New York: Random House.

Zuboff, S. (1988), *In the Age of the Smart Machine*, Oxford: Heinemann.

12. Changing Identities of European Farmers

Marjoleine Hennis

For many years, European farmers have been known for their influential collective action in agricultural policy-making. This collective action was based on national farmers' organizations which represented all types of farmers. Its success was to a large extent connected to the monopolist position of these organizations within the Member States and their strong connections with the national governments. Recently, however, important changes in this pattern have occurred.

In this chapter the link between these changes and the ongoing process of globalization is analysed. The analysis is based on the assumption that globalization affects both the conditions for production and regulation. Globalization is understood as the spread of supraterritorial or transborder relations within the capitalist economy, that is, the causes and consequences of globalization are bound up with the dynamics of surplus of accumulation and the workings of capitalism, including in particular the activities of the state (Scholte 1997, pp. 428–32). Globalization is not understood as an a-political process characterized simply by the adaptation to technologically driven change leading inevitably to the strengthening of market forces. Although this is considered to be an element of the process, it is expected that new forms of opposition will contribute to the outcome of the process of globalization (McMichael 1997, p. 645).

This chapter will focus on the way in which the forces of globalization have affected the identities of farmers and their organizations, and how their ways of coping with the changes have resulted in a new power configuration and parallel agricultural policy. In the first and second sections the impact of globalization and changes in regulation in European agriculture is described. The third section deals with the impact of these changes upon the role of farmers within the food chain and in agricultural policy-making. Section four shows how the identity of farmers has changed in reaction to the forces of globalization and changes in regulation.

GLOBALIZATION

The most recent effects of globalization in European agriculture are based on a process of agricultural restructuring which started in the 1970s. This process consisted of important changes within the agro-industrial production chain.[1] While a shift from labour-intensive to capital-intensive production and flexible forms of labour organization took place, the connections between farmers and supplying and processing industries increased. Industrialization subordinated farms to emerging agro-food corporations, both as buyers of inputs and providers of materials for food manufacturing. The role of agriculture as a whole changed: because of the integration within the agro-industrial production chain, agricultural products changed from final to intermediate products. In this way, parts of agriculture turned into commodity production on the basis of wage labour as mass production of durable goods required standard agricultural raw materials. Enlarged agro-industries engaged in more fixed market commitments with chain stores. In turn, they needed to increase their control over the quantity and timing of their transactions with farm producers (Koning 1995, p. 11). In order to assure themselves of a stable input, corporations introduced contracting with specialized and standardized farms. In spite of the reduction in decision-making and autonomy these contracts brought to farmers, they increasingly engaged in it as a means of acquiring external capital and resources. In general, these processes created a greater homogeneity in strategy between the vertically organized segments of a commodity chain and reduced the choice of farmers.

Biotechnology has come to play an important role in this process. New techniques which break down field crops into generic intermediate food ingredients for use in downstream processing have made the close identification with the specific crop redundant, and have led to a situation in which agricultural inputs give way to non-food crops for the food industry. This process, which is also denoted by Goodman (1991) as *substitutionism*, signifies the uncoupling of agricultural production from food processing, and hence a greater independence of processing industries from a specific input. This is likely to succeed the present, widely used practices of replacing ‘traditional’ inputs (such as wheat for fodder), with tapioca and soya, or sugar and sweeteners on a global scale (*global sourcing*). With these technological innovations, together with ongoing processes of concentration, the flexibility of processing industries has increased considerably and has strengthened their position with respect to primary producers.

The development of European agriculture has depended heavily on capital provided by banks. Not only have banks supplied credit to primary producers for investments in new technology and equipment, they have also facilitated trade taking place on a wide geographical basis, and have fulfilled the function of financial intermediaries through which they could dominate the availability of capital. With the supply of capital, banks have played an important role in the evolution from labour-intensive to capital-intensive production. Moreover, they have considerably influenced the direction of capital investment through their selection of the (type) of farmers who, according to their criteria, were creditworthy. In this way, banks have been able to control substantially the type, rate and direction of capital intensification in the agricultural structure which has contributed to the disparities among farmers. In addition, the credit function of banks has not only strengthened their own position, it has also played a central role in the consolidation of the industrial sector during its formative stage (Boyd 1995). The impact of the huge supply of credit by banks has also been considerable with regard to the displacement of farmers out of production. By facilitating overproduction (with easy credit), prices decreased and forced independent small producers (in all parts of the food chain) in particular into a cost-price squeeze, and ultimately out of business (Boyd 1995).

The restructuring process in European agriculture has changed the characteristics of farms in Europe. Farms have become larger, fewer in number and more specialized. In France, for example, between 1970 and 1995, farm size has increased from 18.9 ha to 38.5 ha (*Agra Europe* 1997). A major consequence of the trend towards fewer and larger farm holdings has been the remarkable loss of farms and farm labourers. In addition, the restructuring process has led to an enormous increase in productivity and agricultural intensification. Wheat, maize, grain, poultry and pigs, especially, are more and more farmed under capital-intensive, factory-like systems where livestock converts grains into a meat product.

The process of globalization has also influenced the relations within the food chain. Leading chemical firms have strengthened their position internationally and in the food chain by mergers and acquisitions. In this way, 'crop development conglomerates' have developed in which the transnational chemical firm maintains links, not only with one or more plant breeding firms, but also with software firms and public research institutes through takeovers or through collaboration (Pistorius and van Wijk 1999, p. 118). Also, processing industries have started to diversify their activities within the food chain in order to maintain control over the rest of the food chain (Goodman 1991, p. 44). Nestlé and Unilever, for example, do much of the distributing themselves, and Ferruzzi is not only involved in the trade of cereals and sugar, but has also commenced activities in the chemical industry by purchasing Montedison. This process of concentration has created a greater homogeneity in strategy between the vertically organized segments of commodity chains.

Although the relations within the food chain remain to a large extent dominated by the upstream part, based on input-orientated corporate capital, they are increasingly associated with the control and construction of value from the point of consumption. This empowers near-consumer agencies. In other words, retailers have come to play a more and more significant role within the food chain as the quality of food has gained in importance. Given the greater role of the demand side on the food market, retailers are in a position to impose a certain mode of production by establishing conditions and quality standards for products. With this development, a shift can be noticed from quantitatively regulated food policies to qualitative ones. That is, the basis of social action within the food chain has tended to move from control on inputs and standardization to quality, regulation and consumption (Marsden 1995).

In sum, the development towards the global organization of production has restructured European agricultural production and has reorganized the relations among production forces within the agro-industrial production chain. Processing industries and retailers have gained a relatively powerful position with respect to agricultural producers. As part of this, the relations among farmers, the concept of farming as such, and its role in the policy-making process have undergone changes in the EU. The process has been enhanced by agricultural policy and has led to a general pattern of change in European agriculture.

REGULATION

The restructuring of European agriculture has been enhanced by the Common Agricultural Policy (CAP). The most important pillar of the CAP, the market and price policy, was designed to provide farmers with a decent income and consumers with a secure food supply. Until the early 1990s, it made up for two-thirds of the EU budget. The idea behind this policy was that producers should receive their income through the price for their products. In this way, the CAP transferred the income from consumers to producers through high prices of agricultural products. Thus, net-importing Member States for food transferred income to net-exporting countries through trade and through the Common budget. This mechanism encouraged Member States to increase domestic production either in order to maximize net recipients or to minimize their net contribution to the budget. Since the support went mainly to commodities produced on relatively modernized holdings, it soon led to surpluses. The policy gave also an incentive to the increase of farm scale and to increased use of capital through the provision of high and stable prices.

Regulation has changed under the process of globalization. The globalization of capital has had an impact on the state through its function of mediating international capital and through the wider social processes connected to it. Two general tendencies in regulation can be distinguished: on the one hand, in the industrialized world there has been a dismantling of state controlled or sponsored structures for the planning of agriculture, and a move towards more market-based forms of regulation (market deregulation). The reform of the CAP in 1992 and the Uruguay Round Agreement on Agriculture (URAA) illustrate this. The major change in agricultural regulation emerges from the altered role of the state in the global economy, and from the difficulties occasioned by an expansion of national agricultural output which has outstripped domestic markets as well as the capacity of natural systems to absorb the increase (Lowe, Marsden and Whatmore 1994). On the other hand, the implementation of deregulatory measures has often been accompanied by reregulation. On the one hand, this has been a way to make up for the process of deregulation. The efforts to control the costs of the CAP through measures such as dairy quotas and set-aside, which were continued with the partial deregulation of the CAP in 1992 (the MacSharry reforms), and with the URAA were all accompanied by a host of exemptions, safeguards, and sideline measures to protect domestic commodity interests (Lowe, Marsden and Whatmore 1994, p. 2). On the other hand, the increased role of the consumer, together with the negative side-effects of the productivist model of agricultural production for the environment have negatively affected the image of agriculture in the public opinion which is reflected in regulation.

The link between a greater concern for the environment and food safety, and agriculture has grown since the 1970s. During the 1970s and 1980s a wave of environmental mobilization appeared, mainly in the Western world. This is shown from the increase in members of environmental groups at both the local and national level (Symes 1992). An estimate made in 1985 held that more than 10 million Europeans belonged to one or more environmental groups, and even more were active in local environmental action groups (Dalton 1993, p. 64). While the first wave of environmental mobilization, which had taken place at the turn of the century, was conservationist in character, the new wave of the 1970s and

1980s focused on issues such as nuclear power, industrial pollution, acid rain and so on (Dalton 1993). The environmental movement today consists of organizations representing both perspectives.

During the same period, the countryside gained importance in the public perception for recreation, landscape amenity and the preservation of traditional cultures and values. It might have been expected that, together with this new perception of the countryside, agricultural producers would be the first to be attacked. Instead, the approach toward agriculture remained, for a long time, conservationist in character. As Harper notes 'there has long been little public perception of agriculture as both a polluter and an environmental hazard, exploiting and destroying the environment in its attempt to respond to the demands of urban modern life, which it sustains' (Harper 1993, p. 6). Furthermore, those members of green or environmentalist groups interested in agriculture or 'nature', often tend to reject in some sense modernity, and attempt to sustain and re-create the idyll more than looking at modern agricultural production (Harper 1993). This misunderstanding of the countryside led ecologists to overlook the links between agriculture and environmental problems, and often resulted in a conservationist tendency of the rural without looking at modern agricultural farming systems.

The influence of environmental orientations culminates in environmental pressure groups and green political parties. With respect to agriculture, in spite of the marginal role which agriculture has played for a long time among those groups, they have contributed to the growth of environmental consciousness within the EU, and have helped to strengthen the role of DG for Environment as EC institutions attempt to respond to these new political forces and public pressure by enacting environmental regulation, as had already happened within many Member States (Vogel 1993, p. 183). Mainly because of its diversity in perspectives and consequently the variety in ways of approaching the policy process, there has been a rapid integration of environmental issues in the policy process during the 1980s. The different ideological views underlying this diversity have pressed the government for policy reform on several fronts (Dalton 1993, p. 64). Apart from environmental regulation affecting agricultural production, it has had repercussions in a more indirect way, too, for the CAP and its openness to integrating environmental issues.

Together with growing concern of environmental concerns with regard to agriculture, and also in direct reaction to the results of the restructuring process, consumers have increasingly paid attention to the safety of food. As incomes have risen and consumption patterns have changed, consumers have substituted higher value added processed foods for unprocessed foods. They have moved up the food chain. As a result, a greater proportion of the price paid by consumers for final products is accounted for by the food manufacturing and processing industries. The structure of the industry has also changed quite considerably. In particular, concentration on food manufacturing and retailing has increased. Retailing and processing industries are very sensitive to consumers' demands and seek to react as soon as possible to changes in taste. Food suppliers and retailers have begun to realize that a green image may be good for business.

The growing demand for localized and crafted products of high quality and identifiable origin, especially among well-off and concerned consumers, has given an incentive to retailers and processing industries to give more importance to these aspects (Murdoch, Ward and Lowe 1992). Furthermore, as demand for food is rather inelastic, while supply has only increased during the last decades, differentiation of products has become an important strategy among processing industries and retailers in order to continue attracting consumers. The supply of organic products is one way to differentiate. In turn, these retailers, under pressure from their customers and the public opinion in general, have started to demand from processors and agricultural producers that they apply and evaluate the environmental implications of their products, so that they can sell a product which has been produced in a responsible way.

The role of consumers has thus improved within the food chain with regard to environmental and food safety standards in the production process. Via processing industries and retailers, consumers have put pressure on farmers and their mode of production. This usually takes place through contract production and not so much through environmental regulation.[2] European retailers, for example, are shown to attribute an ever more important role to the demands of consumers. The biggest retailers especially are able to impose a quality standard on food producers through contracts. A good example is Europ-Gap, a regime of Good Agricultural Practice which went into force in June 2001. Food producers who do not comply with the standards set in Europ-Gap will not be able to distribute their goods through the 30 European supermarket chains which have signed it.

The position of consumers in agricultural policy-making has considerably been strengthened during the second half of the 1990s. In that period, in direct reaction to major food scandals in Europe, ranging from dioxins in pig feed in Belgium, sludge in feed in France, to the much more diffused Bovine Spongiform Encephalopathy (BSE) or 'mad cow disease',[3] consumer concerns shifted towards food safety issues. The apparent negligence of food safety aspects and the failure to control the actors in the food chain by national and European authorities at the start of the crisis led to a dramatic decrease in consumer confidence and to corresponding consumption patterns and collective action. While the consumption of beef in certain regions decreased (for limited periods of time) by 80 per cent, campaigns by consumer organizations against farm practices that neglected animal well-being, human health risks and the environment, intensified. The agricultural model of production as enhanced under the CAP came widely to be perceived as the major cause of these problems. The impact of the crises was so serious that important changes were set in motion in the European policy process and have contributed to the trend towards re-regulation.

In reaction to the food scandals, article 153 of the Amsterdam Treaty (1997) was adopted which called upon all other EU policies to integrate consumer policies. This meant that consumer policy became an integral part of policy development across all Directorate Generals (DGs), as well as in the European Parliament (EP) and in the Council. For the CAP the integration of consumer policies meant a greater emphasis on quality products, which on the one hand, led the CAP to come more into line with the already existing practices in the food chain. On the other hand, this integration led to increased re-regulation.

A little later, the European Commission issued the White Paper on Food Safety (12 January 2000) in which it identified food safety as one of its top priorities. The White Paper set out the plans for a proactive new food policy which aimed at a major control from 'farm to fork'. The establishment of the European Food Authority had to safeguard the continuance of these policy objectives. Along the same lines, the Member States adopted Council Directive 2000/13/EC on labelling and traceability in order to make the food chain more transparent and to provide the consumer with more information. In combination with article 153 of the Treaty of Amsterdam, these measures led to a greater involvement in food production by DGs other than the DG for Agriculture, especially by the DG for Consumers and Consumer health protection (DG SANCO). In fact, Food Safety issues were shifted from DG for Agriculture to the latter.

Thus, the role of consumption has increased relative to that of production within the agro-industrial production chain, while consumers have grown more sensitive to the effects of the industrial mode of production and to its additives to food in growing, storing, and processing stages. As a result, food production is increasingly regulated in the form of labelling and other legislation. It has been reinforced by retailers who have generated strong pressures for adapting (and not abolishing) regulation to their advantage. These developments are reflected in an increase of environmental and food safety regulation and have forced farmers to adapt to the new situation.

FARMERS LOSING GROUND

The restructuring process has had a homogenizing effect in the sense that it has created a system of mass-production based on standardized production methods on a large scale. At the same time, it has led to increasing differentiation between those farmers who have integrated with the rest of the food chain, and those who have not. In other words, not all farmers have succeeded in the adoption of the industrial production model, nor in some cases have they wished to adopt it, for example, in the form of contractual arrangements with agro-industries.

In general, relatively modern, commercial and large-scale farms have been in the best position to respond to the demands of processors and distributors. Small-scale, labour-intensive farms have mostly been left out. In this way, a division has developed between one group of farmers, which has been attractive enough for agro-industry to receive contracts and to 'quasi-integrate' with agro-industry (Koning 1995, p. 11) and another group of farmers which has not integrated with agro-industry at all. In the literature this development is also referred to as a *two-track* agricultural system, in which large farmers, using industrial production methods, account for the bulk of food production and operate mainly on the market, while smaller producers are more focused on regional markets and are often kept in farming through environmental grants (Clunies-Ross and Hildyard 1992).

The transformation of the agricultural structure and the changes in regulation have formed a continuous pressure on farmers. For example, the market and price policy of the CAP has had a far-reaching impact on the way agricultural production has developed in Europe since the 1960s. However, after a long period of state intervention, farmers are now increasingly confronted with deregulation. Deregulation means for farmers (with some exceptions) that governmental interventions in export markets will diminish as prices approach international levels. Ongoing market deregulation has increased competition among farmers as processing industries have obtained a wider choice of inputs. Mainly large-scale producers who have benefited disproportionately from price support programs, and transnational corporations (TNCs), involved in the food industry which would be able to further centralize economic power, have gained (McMichael 1994, p. 180). These large-scale farmers have been able to stay competitive by investing in the latest technology, whereas the remaining group of farmers can only opt for a strategy based on the lowest possible capital expenses, together with an off-farm job in order to achieve a reasonable income (Almas 1994, p. 22). Thus, the trend towards deregulation is welcomed only by a limited group of farmers together with agro-industries. This has contributed to the problems of collective action among farmers and to the development of diverging identities of agricultural producers who used to be organized within the same organization.

Another group of farmers sees more opportunities in re-regulation of agriculture, although this has created greater restrictions on farming. Possibilities for farmers are found in dropping the conventional model of agricultural production which has traditionally been imposed by the CAP. For example, production could be tied to the assertion of positive environmental images of place, thus assisting local produce to carve out niche markets (Murdoch, Ward and Lowe 1992 p. 14). By shortening the food chain and by keeping most activities on the farm and serving regional markets, the additional advantage for farmers is their decreasing dependence on processing industries and retailers. Most importantly perhaps, these farmers benefit from the large variety of subsidies or agri-environmental measures,[4] at a time in which financial aid for conventional agriculture declines. Examples of agri-environmental measures are Regulation no. 2080/92, a Community aid scheme for forestry measures in agriculture, and Regulation 2078/92, which aims to encourage farmers to adopt environmentally sound production methods, or to take steps to safeguard nature and the landscape. Aid is fixed as a function of the area concerned and the type of crop to be produced. In 1995, the Commission had already approved 160 agri-environmental programmes submitted by the Member States or the regions under the latter regulation.

In order to increase their choices, farmers seek to augment their influence on agricultural policy through membership in alternative interest organizations, such as alliances between environmental groups and farmers and consumers. These alliances play an important role in the strengthening of an integrated approach which counterbalances the dominant modernization model in the CAP that for a long time has favoured an elite group of farmers.

Thus, for farmers the picture has become complex. In an attempt to move away from the price support system, and at the same time adopting an environmental approach, the CAP has been reformed. The result is a mixture of principles linked to both the conventional type of agricultural production and to production-restrictive measures with a beneficial side-effect for the environment and rural development, and food safety. Under pressure from production-enhancing policies, farmers still tend to give more importance to production strategies and food production than to environmental goods. At the same time, however, the traditional voluntary character of environmental conservation has given way to limitations on farm activities imposed by environmental regulations or considerations about the landscape. Those regulations mean that the rights of the farmer on his or her land are gradually being made subordinate to the wider public benefit. Thus, with the introduction of environmental and food safety issues the nature of CAP intervention has changed. At the farm level this means a direct limitation of the freedom of farmers in their production activities.

CHANGING IDENTITIES

The increasing differentiation among farmers has not only affected their role within the food chain. The differentiation has also made it increasingly difficult for farmers' organizations to continue to legitimate the traditionally general representation of farmers, while their influence on policy-making has declined. In addition to their diminished capacity to regulate their own constituencies and reach bargained agreements with state agencies, the traditional organizations have been confronted with the rise of new ones. Furthermore, the opposite tendencies in regulation – market deregulation and environmental and food safety re-regulation – combined with a growing criticism of the problems of surpluses of agricultural production have led to political and institutional tensions in the corporatist mode of capital-intensive production between the regulation of global agro-food trade on the one hand, and regional and national farm support on the other hand. Farmers have become increasingly aware of these changes and have sought to react according to their possibilities within the food chain. Similar patterns have occurred in most European Member States.

In the Netherlands, for example, a situation in which three general farmers' organizations based upon socio-religious principles represented all interests in Dutch agriculture came under siege during the mid-1990s. In 1994, a national debate among members of farmers' organizations, politicians, environmental groups and consumers was initiated. The aim was to provoke a nation-wide discussion about the future of agriculture, without the organizational structure which had developed around Dutch agricultural policy-making and which was increasingly perceived as limiting. The initiative for, and the relative success of the debate showed that two important changes had taken place. First, it underlined a change in the attitude among the members of the agricultural sector. In the debate it was publicly acknowledged, at all levels of the agrarian community, that the way agriculture was operating had (partially) failed. Secondly, the fact that the discussion was conducted outside the protected structure of agricultural interest intermediation illustrated that its limitations in handling a process of profound change were felt. The debate, propelled by the crisis connected to the restructuring process in agriculture and changes in regulation has been indicative for the re-definition of the role of Dutch farmers in food production and in policy-making in 1995.

In France, a similar development took place. At the beginning of the 1990s, against the background of the CAP reform and the URAA, the waning influence of the major farmers' organization in policy-making was reflected in frequent protest actions by farmers in the streets of Paris, or less pacifically, by farmers obstructing the entrance of agricultural goods from other Member States into the country. This was only a prelude to the changes within French farm politics. In a survey, held in 1997 among French farmers about their expectations concerning the future of farming (FNPAR 1997), mainly two types of reactions were recorded. On the one hand, large-scale producers in the regions of the North of France and the Basin showed a predominantly defensive attitude towards changes in national and European policy. On the other hand, small farmers and producers of quality products focused on the possibilities of the market and the reform of agricultural policy with respect to quality products, organic farming, and other functions of agriculture (*BIMA* 1998, p. 37). Significantly, groups of similar size expressed a preference for a general farmers' organization (19.7 per cent) or a vertical organization (19.3 per cent) (*BIMA* 1998, p. 37). Those divergences affected the field of interest representation. For example, the largest (and general) farmers' organization, FNSEA, that focused traditionally on the productivist model of farming, lost members to alternative organizations which paid more attention to small farmers, such as the Confédération Paysanne. These developments have grown stronger from the mid-1990s on, and have resulted in important changes in interest representation and a new agricultural policy which reflects the new multiple identities of farmers.

Also in other European Member States this kind of development has taken place. For example, in Germany many new organizations, which focus on alternative farming practices or small farmers, have seen a significant increase in members since the 1990s. Moreover, farmers have lost their influence in agricultural policy-making. National farm policies have become more open to a greater number of actors outside the direct realm of food production. For example, under direct pressure of the BSE crisis, the German Ministry of Agriculture has come to include consumer protection, and the British Ministry of Agriculture, Fisheries, and Food (MAFF) was merged in June 2001 with rural and environmental policy to create DEFRA.

In sum, the greater differentiation among farmers leading to problems of legitimacy of traditional farmers' organizations and less influence in farm policies, together with a changing perception of agriculture in the public opinion, have forced farmers to shift away from conventional schemes of production and representation and to search for new identities of farmers as strategies for survival.

Agricultural producers who have successfully integrated with agro-industry have reacted differently from small producers. Large-scale, commercial farmers continue to focus on food production as their main activity, although quality production is increasingly taken into account. Their integration with agro-industry has made their interests closer to those of the industrial sector. This group of farmers tends to present itself, not anymore as farmers who depend on financial support by the state, but as market-orientated entrepreneurs who demand that farmers' organizations focus their activities on improving the relations within the food chain rather than those with policy-makers, and demand state intervention only for what regards a better functioning of the market. In terms of collective action, these farmers are increasingly successful through vertical commodity organizations in which all actors in the food chain for a certain commodity are represented.

At the other end of the scale are farmers who remain more dependent on state intervention. In order to continue receiving financial aid, in a time of decreasing price support and more attention for the environment and quality products, some of these farmers have taken initiatives to shorten the food chain and to produce directly for the market. Others have turned to organic methods of farming in reaction to the increasing dominance of agro-industry within the food chain. These initiatives have been accompanied by the growing need of retailers for differentiation in food products as they try to widen the spectrum of products adapted to different consumer groups. In general, these farmers have increasingly shifted from the focus on food production alone towards a new identity in which they fulfil various functions with regard to rural environment. Their new activities, besides food production, include the management of the landscape, the safeguarding of the environment, tourism and the provision of rural employment. These farmers receive their income, not only through the market, but continue to depend on subsidies, for example in the form of territorial contracts with the state through which they are paid for the various services they offer. They find a basis for collective action in alliances with farm workers, consumer groups and environmentalists.

The new identities, which have developed as strategies for survival, are increasingly reflected in interest representation and agricultural policy in Europe. Over the past few years, national and European agricultural policies have come to be based upon a model of interest intermediation consisting of a much greater number of actors, and translated into a much greater number of functions related to agriculture. For example, in the Netherlands, in reaction to the crisis of legitimacy of the three farmers' organizations in the beginning of the 1990s, a more decentralized model of interest-representation under the name LTO-NL, in which the various vertical groups lobbied the Dutch Ministry of Agriculture, took their place. Besides organizational changes, LTO worked on the development of a new common denominator among farmers, which had to take account of the structural changes in the sector, of the changes in regulation and, most importantly, which was able to go beyond the increasing differences among farmers. Vision reports with titles such as 'Harvesting in a new century' (1999) and 'Strengthening through change' (2000) in which these new strategies were discussed, illustrated the need to build on a new, collective identity of farmers. It has resulted in a mission statement of LTO which is based on the premise that the *licence to produce* provided by society to farmers is in danger and that this problem must be dealt with by shifting towards sustainable agricultural production, both for market-orientated farmers and for farmers who offer other services to society such as, for example, landscape management (LTO 2001). In practice, this strategy has not been able to conceal the increasing gap between entrepreneurial farmers and service-orientated farmers. According to LTO chairman, Doornbos, this situation forces the LTO to make a choice which is likely to result in a predominantly market-orientated approach, and a shift from its strong linkages with policy-makers towards increased partnerships with the other actors in the food chain (*Financieel Dagblad* 2001).

So far, Dutch agricultural policy-making has undergone some changes that reflect the tensions in agriculture. A good example is an increase in the number of actors coming from outside agricultural production with access to the decision-making process. It is reflected in yet another national debate, organized in 2001, about a hot issue regarding agriculture and food, that is genetically modified organisms (GMOs). This debate which, this time, has been a joint initiative of the Ministry of Agriculture and the Ministry of Public Health and Environment, shows the changing climate in agricultural policy-making by making consumer acceptance a priority.

Agricultural policy focuses on Dutch competitiveness through national champions, rather than on the support of the sector in general. Furthermore, in order to achieve greater international competitiveness, the policy pays more attention to other segments in the food chain, such as processing industries and retailers. The most important instruments in this regard are tax cuts and deregulation. The increased emphasis on the environment seems part of the focus on the increased competitiveness of agriculture. Sustainable agriculture has become a market strategy which responds to the increased demands of consumers for quality goods. This rather limited view on organic farming is reflected in the way it is stimulated through the policy. State intervention in the form of the promotion of research in better technologies and innovation should enhance Dutch organic farming.[5]

Another clear example of the development towards a more differentiated situation in agriculture is France. First, more actors have entered to the policy-making process. Already in the mid-1990s, the alternative organization, Confédération Paysanne, was granted access to the policy-making process after it received more than 20 per cent of the votes in the policy-making in the *Chambres d'Agriculture* (Agricultural Chambers), a semi-public body. In the elections of 2001 the position of the Confédération Paysanne was reinforced when it received 26 per cent of the vote. This has disrupted the system in which FNSEA (together with CNJA, the young farmers' organization) was the only one to have access to the agricultural policy-makers. Secondly, the greater number of actors is reflected in a new agricultural policy. In January 2000, the *Loi d'Orientation*, a guidance law, came into effect. This law underlines the attention in French agricultural policy-making for the dualistic features of farming, that is, to maintain the strong export position of France, and the role of farmers as guardians of the landscape and the environment. Its most important feature is the introduction of the *Contrat Territorial d'Exploitation* (CTE), the aforementioned contract between the state and the farmer for the provision of various functions with regard to the rural environment. At the same time, the guidance law has led to a greater openness towards vertically organized commodity groups. The vertical organizations are recognized and consulted by the state about agricultural policy regarding food chains. In addition, they are authorized to make decisions about certain directives of national policy (*BIMA* 1998, p. 12).

More recently, initiatives have also been taken in Italy in order to respond to changes in the food chain and policy-making. The major Italian farmers' organization, Coldiretti, in 2001 organized a debate to which it invited all actors of the food chain and policy-makers in order to collectively develop a strategy in response to greater consumer concern.

One tangible result so far has been the *Patto con il consumatore* (deal with the consumer) which consists in a commitment by those farmers who adhere to Coldiretti to provide more information about the production process to consumers, and to prioritize the quality of food. It underlines the changed identity of the farmers from contributors to food security, to a clear responsibility also for the comfort and the health of consumers, as well as for the environment. Traceability is one of the instruments through which this is to be achieved.

At the European level, the DG for Agriculture has done everything to keep pace with the changes. The new identities of farmers are reflected in the new plan for reform of the CAP under Agenda 2000 as agreed upon by the heads of the Member States in Berlin, March 1999 and confirmed by the Commission in July 2002, in its report 'Towards sustainable farming' and presented at the Mid-Term Review of the EU's Common Agricultural Policy.

On the one hand, Agenda 2000 focuses upon the maintenance of the competitiveness and export potential of European agriculture while complying with WTO regulation. This implies a movement towards world prices, a reduction of intervention levels, and a greater decentralization of policy implementation. On the other hand, Agenda 2000, has been a response to the demands for a shift away from a focus on food security to more attention for food safety. The new CAP is based on two pillars: the first pillar focuses on policies for agricultural markets, and the second pillar on policies for rural development. It means that, alongside the measures which focus on the competitiveness of European agriculture, more attention is given to the development of rural areas, thereby taking into account the requirements of consumers and environmental issues. In other words, the second group of policies aims at enhancing employment, the creation of new sources of income (for example, through agri-tourism), and the protection of rural heritage. The second pillar is based on various principles, of which the most important is the principle of multifunctionality of agriculture. Multifunctionality means that various services provided by farmers, other than food production, are encouraged. These services include for example the management of the landscape.

The ultimate objective of the CAP reform has been the integration of environmental and food safety issues. The fact that price and market policies (which may include the enhancement of quality products) on the one hand, and rural development through the principle of multifunctionality on the other hand, which aim more clearly on environmental and food safety requirements remain clearly separated shows the difficulty to achieve a truly integrated approach. Nevertheless, this development has given way for an increasing number of new actors into agricultural policy-making and, therefore, continues to put pressure on farmers.

CONCLUSION

This chapter has shown the way in which the forces of globalization have led to a greater degree of differentiation in the roles of farmers within the agro-industrial production chain. Parallel to these changes at the production level, the identities of farmers have changed which has forced them to change also the way they used to organize their interests. Whereas some farmers have increasingly identified with the interests of agro-industries and have tended to vertical organizations of interests, others have opted for creating a greater distance with other parts of the agro-industrial production chain while producing more directly for the market or depending more on alternative support systems.

The differentiation in identities among farmers, together with an increasing number of actors from outside agricultural production with an interest in agricultural policy-making, such as consumers and environmentalists, puts more and more pressure on agricultural policy. Although still brought together under one agricultural policy, the new identities of farmers seem increasingly to diverge. In a way, this is enhanced by the policies that reward, in the first place, farmers that are already performing well and are able to perform many different functions and furthermore, by underlining the differences between the different potentialities of farmers. It is gradually leading to a situation in which one group of farmers comes to depend largely on the financial rewards for the management of landscapes, whereas others obtain their income mainly through food production and have more and more interests in common with the industrial sector. In this way, a situation may occur in which the new identities will increasingly replace the old identity of 'farmer' which will make it more and more difficult to manage both farmers' organizations and agricultural policies, and may eventually lead to their disappearance.

NOTES

1. The term agro-industrial production chain refers to the sum of all sectors, which together are responsible for an agricultural product to end up in a shop for final consumption. It consists basically of four phases of production (Ruivenkamp 1989): suppliers (of seeds, pesticides, machinery and so on); agricultural production; processing; and distribution or retailing.

2. In these cases, the corporations which give the contracts in general prefer environmental regulation based on the polluter-pays principle. In this way, they can demand certain standards and impose a certain mode of production without having to pay for the environmental damage caused by it.
3. This disease is caused by the use of performance-enhancing dairy feed. Its proteins come from meat and bone meal from slaughtered sheep and cows. In 1984, this disease first appeared in a cow in Britain. Ten years later, it was discovered that the disease could also spread to humans by eating parts of the nerve system and lymphs of cows. At the beginning of 2002, however, on the basis of experiments with mice, the Nobel Prize Winner, Prusiner, who discovered the protein that caused the human version of BSE, concluded that there might also be a risk for humans by eating the other parts of the cow.
4. Agri-environmental measures are 'policies whereby farmers ... are offered incentives which are intended to influence management decisions, and to result in a desired environmental outcome' (Baldock, 1996). The measures are usually voluntary for farmers in the sense that they are free to enrol in such a programme. Sometimes, however, they are 'quasi-voluntary', for example, when farmers are required to enrol in an environmental programme in order to qualify for conventional production support.
5. This policy has not been very successful. Whereas the number of organic farms in 1999 had increased to 25 per cent, in 2001 this had fallen to 8 per cent, while approximately half a million Euros remain unused. Biologica, the umbrella organization of organic farmers attributes this decline to confusing regulation and the general shrinking of agriculture.

REFERENCES

Agra Europe (1997), London.

Almas, R. (1994), 'The rise and fall of agricultural policy cycles: from planned economy to green liberalism', *Journal of Rural Studies*, **10** (1), 15–25.

Baldock, D. (1996), 'Environmental impacts of agri-environmental measures', in *Subsidies and Environment, Exploring Linkages*, OECD document, Paris: OECD.

BIMA (1998), no. 1475, Paris: Ministère de l'Agriculture et de la Peche.

Boyd, W. (1995), 'Agrarian transitions, agro-industrial restructuring, and the politics of flexibility in the American South', paper prepared for workshop on the Political Economy of the Agro-Food System, Berkeley: Institute of International Studies, University of California, September.

Clunies-Ross, T. and Hildyard, N. (1992), *The Politics of Industrial Agriculture*, London: Earthscan.

Dalton, R. J. (1993), 'The environmental movement in Western Europe', in S. Kamieniecki (ed.), *Environmental Politics in the Internation Arena:*

Movements, Parties, Organizations and Policy, Albany: State University of New York Press, pp. 41–68.

European Commission (2002), 'Towards sustainable farming', IP/02/1026, Brussels: European Commission.

Financieel Dagblad (2001), 'Ondernemende boer zaait tweespalt', 11 December.

FNPAR (1997), 'Sondage: "Les Français font confiance aux agriculteurs"', *La Propriété Agricole*, no. 2.

Goodman, D. (1991) 'Some recent tendencies in the industrial reorganization of the agri-food system' in W. H. Friedland, L. Basch, F. H. Buttel and A. P. Rady (eds), *Towards a New Political Economy of Agriculture*, Boulder/San Francisco/Oxford: Westview Press, pp. 37–66.

Harper, S. (ed.) (1993), *The Greening of Rural Policy, International Perspectives*, London and New York: Belhaven Press.

Koning, N. (1995), 'Fordism, "post-Fordism", and the agrarian question', paper prepared for conference on Agrarian Questions, Wageningen, 22–24 May.

Lowe, P., T. Marsden and S. Whatmore (eds) (1994), *Regulating Agriculture*, London: David Fulton.

LTO (2001), 'Oogsten in een nieuwe eeuw, toetsstenen voor het jaarplan 2002 van het LTO concern', Den Haag: LTO, July.

Marsden, T. (1995), 'The peculiarity and distinctiveness of recent agrarian development: dominant and dependent spaces', paper prepared for workshop on the Political Economy of the Agro-Food System, Berkeley: Institute of International Studies, University of California, September.

McMichael, P. (1994), 'GATT, global regulation and the construction of a new hegemonic order', in P. Lowe, T. Marsden and S. Whatmore (eds), *Regulating Agriculture*, London: David Fulton, pp. 163–90.

McMichael, P. (1997), 'Rethinking globalization: the agrarian question revisited', *Review of International Political Economy*, **4** (4), 630–62.

Murdoch, J., N. Ward and P. Lowe (1992), 'Rural sustainable development: a socio-political perspective on the role of agriculture', paper presented at the Circle for Rural European Studies Conference, Crete: Mediterranean Agronomic Institute of Chania, October.

Pistorius, R. and J. van Wijk (1999), 'The exploitation of plant genetic information, political strategies in crop development', PhD thesis, Amsterdam: University of Amsterdam.

Ruivenkamp, G. (1989), *De invoering van biotechnologie in de agro-industriële productieketen*, Utrecht: Jan van Arkel.

Scholte, J. A. (1997), 'Global capitalism and the state', *International Affairs*, **73** (3), 427–52.

Symes, D. (1992), 'Agriculture, the state and rural society in Europe: trends and issues', *Sociologia Ruralis*, **32** (2/3), 193–208.

Vogel, D. (1993), 'Environmental policy in the European Community', in S. Kamieniecki (ed.), *Environmental Politics in the International Arena: Movements, Parties, Organizations and Policy*, Albany: State University of New York Press, pp. 203–22.

13. Biotechnology: The Production of New Identities

Guido Ruivenkamp

INTRODUCTION

> I never thought that I would live through a revolution, but I believe that there is one going on. No Winter Palace has been seized, no Bastille stormed, no monarchy abolished. Political orders around the world remain alas, very much the same. But on the industrial and scientific fronts major changes are gathering momentum. A technological assault is being prepared that will transform the economies of developed and developing nations. Its substance is the engineering of life processes for commercial ends: biotechnology.
>
> (Yoxen 1983, p. 14)[1]

Technological developments and globalization are strongly interrelated. First, technological developments can *stimulate the contrasting processes of homogenization and heterogenization*, being the two basic elements of global connectivity. For example, plant breeding technologies can contribute to the *uniformity* of agricultural production in different regions of the world by the development and use of high-yielding varieties. At the same time, however, plant breeding technologies can increase the *diversity* of local farming styles (Van der Ploeg 1991). Secondly, the homogenization and heterogenization can also be *reflected in the development of technologies*. For example, the impressive growth of herbicide- and insect-tolerant crops in the world – from 1.7 million hectares in 1996 to 52.6 million hectares in 2001 (James 2001, p. 2) – is a clear illustration of how biotechnology reflects homogenizing developments. On the other hand, at local level one sees different attempts to make a mixture or 'creolization' (Hannerz 1992) of science-based and farmers' field-based laboratories. Therefore, biotechnological developments do not *necessarily* lead to an increasing homogenization and concentration of transnational economic and political power. Simultaneously another set of radically divergent processes takes place. There is evidence that the same individuals (farmers, consumers, scientists), who are subject to the homogenizing developments, are at the same time becoming increasingly *reflexive* with respect to these developments (Beck, Giddens and Lash 1994; Lash and Urry 1994); reflexive in the sense of confronting the exogenous biotechnological developments with possibilities to tailor biotechnologies to endogenous developments (Ruivenkamp 2001).

The aim of this chapter is to conceptualize biotechnology from this dialectic perspective and to understand what role biotechnology plays in re-organizing production and everyday life. Especially, we intend to envisage the ways in which biotechnologies could be used against the dominant interests and relations and become even catalysts for a local sociology of reflexivity (Lash and Urry 1994, p. 5).

In the first section we describe how agro-biotechnology reflects and reinforces homogenizing tendencies in the social organization of global food chains. We describe how biotechnology incorporates two (historical) processes of separating modern agricultural production from its local specificities and contributes to a reorganization of food production (Ruivenkamp 1989). However, we emphasize that biotechnological developments not only change the local 'landscapes' (Appadurai 1990, p. 296) of agriculture and food production, but also the personal contexts of various actors. Biotechnology is emptying out the traditional meanings of various social categories and we stress that the reorganization process even *challenges* the capacity of new group identity-formation, because most actors subjectively still refer to their traditional social categories as being, for example, scientists, working in public or private research institutions, while objectively they are positioned in a changed social environment.

The duality of social developments and actors' sense-making and reacting is at the centre of this book. We argue, however, that this assumed duality of social developments and actors sense-making and reacting is undermined by the reorganization process. We argue that an impressive *de-synchronization* emerges between the productive reorganization process and the subjective identity formation. The de-synchronization is not primarily an (ideological) limitation at individual level, but has even become a basic *characteristic* of the ongoing transformation process. Exactly through the discrepancy between the subjective and objective reconstruction of identities, new forms of political control are emerging. Mechanisms of control that are directly embedded in the reorganization of food production, reflecting a development towards a 'politicization of food production' (Ruivenkamp 1989), in which, for example, scientists and their laboratory products get a direct political identity, while, for example, policy-makers become de-politicized. However, the same process might also create new space enabling the same individuals – subject to the new forms of control – to become increasingly reflexive with respect to these developments and look for a new base on which a subjective reconstruction of identities in relation to the homogenizing and heterogenizing biotechnological developments *may take place*.

The second section indicates that the subjective reconstruction of identities in a politicizing production system can no longer be based on a revival of traditional categories like farmer, scientist and enterpreneur, but might be formed through an active participation in the development of new hybrid forms of socio/biotechnological ensembles at 'glocal level' (Robertson 1995, pp. 25–41). The hypothesis is formulated that actors' sense-making and reacting might increasingly be shaped by their specific contributions to 'creolize' (Hannerz 1992) science- and people-based biotechnologies for endogenous developments (Van der Ploeg and Long 1994). Therefore, we conclude this chapter with a plea to participate in a new international program on tailor-made biotechnologies (Ruivenkamp 2001), set up to search for an equilibrated balance for protecting communities against the homogenizing influences of biotechnological developments and, at the same time, stimulating the integration of heteoregenous eco/biotechnological innovations, attuned to locally orientated developments.

BIOTECHNOLOGY: THE PRODUCTION OF SEPARATIONS

Biotechnology has been defined as 'the application of scientific and engineering principles to the processing of materials by biological agents to provide goods and services' (Bull, Holt and Lilly 1982, p. 21). Since the recombinant-DNA technology became applicable in 1973, vast new possibilities were opened to change and apply different biological agents, as microorganisms, enzymes, plant and animal cells for the commercial manufacturing of a large variety of substances in the pharmaceutical, agriculture and food production sector.

Contemporarily with the scientific breakthroughs in the field of biotechnology in the 1970s a social conflict took place regarding the reorganization of production processes (Crouch and Pizzorno 1977). The rise of biotechnology is partly the result of these economic changes and this political battle (Vergopoulos 1986, p. 1). This conflict regarding new social relations forms the context out of which biotechnological developments took shape (Ruivenkamp 1989). An important social and political change to which the development of biotechnology has been related is the rise of the agro-industrial production chain – a change that illustrates a spatial transformation of production, that is, from geographically concentrated units towards production systems spread over society (Belforte et al. 1976), and a political transformation, that is, a crisis of traditional political identities built up around sectors and the emergence of new social movements (Negri 1979). In correspondence with this spatial and political transformation, also the agricultural sector became increasingly integrated in an international organization of food chains split up into four main sections:

1. The use of inputs such as seed, pesticides and agricultural machinery.
2. The agricultural production.
3. The industrial processing of agricultural products into food stuffs.
4. The international retailing of these manufactured products to consumers.

The already well-established integration of agriculture in the international agro-industrial production chain implies that new technical developments will in the first and third phase of the food chain have a great impact on the social-economic developments in the farming sector. Through the deployment of biotechnology in the production of seed for sowing and agrochemical pesticides the relation between the input-supplying companies and the farmers will change. A new relationship will be created, too, between the farmers and the food-processing industry when biotechnological developments alter the system of processing agricultural products into food products. Within this dynamic process of reorganizing food production, reinforced by specific biotechnological developments, positions and identities of many different actors (farmers, food companies) and actants (seed, biocatalysts, enzymes) are changing. To understand the formation of these new identities, it is necessary first to look at the reorganization process of the food chain. Therefore, we start with describing how the reorganization of the relations between farming and other actors takes place by reinforcing two processes of separations in the food chain.

The Separation of Agriculture from its Ecological Environment

The industrial transformation of agriculture has taken place through two long-term processes: appropriation and substitution (Goodman, Sorj and Wilkinson 1987, cited by Pistorius and van Wijk 1999, p. 17). Appropriation refers to a process in which the controllable elements of farming are being transformed into industrial activities and reintroduced into agriculture as inputs. Substitution refers to processes through which agricultural products are increasingly reduced to industrial semi-products and may eventually be replaced by synthetic products. This process of substitution leads to the uncoupling of agricultural production and consumption (see the next section). The process of appropriation leads to the disconnection of agriculture and its ecological environment. Agro-industrial biotechnological developments reinforce and transform both processes (Ruivenkamp 1989).

Concerning the uncoupling of agriculture from its ecological environment, the development of specific scientific advances, embedded within international competition structures, plays an important role. Through the development of tissue culture, cell fusion and r-DNA techniques, seeds companies can intervene in the genetic structure of plants more radically and efficiently. Originally, the genetic information of a plant was mostly formed in interaction with its environment, which meant that some plants could grow in a temperate and other plants in a warm climate. This strong interrelationship between plants and their environment also implied that only those plants would survive in a specific locality, which had embodied certain characteristics of their environment, like 'their' soil and climate, in their own gene structure. This strong interrelationship between plants and their environment has been reinforced by natural selection processes as well as by farmers' selections, looking for plants 'adapted to their environment'. However, opening up this strong relationship and 'liberating' the plants from their natural constraints have also always been tried. Especially during the Green Revolution, traditional plant breeding techniques have been successful in bringing additional information in the genetic structure of plants. This additional information for higher yields, however, did not always match very well with the plants' incorporated information to protect and survive in their environment. The modified plants became increasingly susceptible to diseases and pests. The modified plants were giving higher yields but also needed more protection 'from outside' to survive in their locality. It is important to stress that these traits 'to survive' were not primarily incorporated in the genetic information package of the new varieties, but in the first instance were added 'from outside' in the form of agro-chemicals. This transformed the interrelationship plant growth–natural environment into a triple relationship plant growth–chemicals–natural environment – a triple relationship that I have called the biochemicalization of agriculture (Ruivenkamp 1989). In view of this new context of plant breeding, it could easily be predicted that biotechnology within this context will lead to a particular group of products.

> Not only new varieties will be developed in interaction with these chemicals, like herbicide-resistant crops. Later on also crops with an in-built protection against insects and diseases will be developed. Even new crops with a changed

> and higher composition of food nutrients and nitrogen-fixation crops will be developed.
>
> (Ruivenkamp 1989, p. 120)

The biochemicalization of agriculture and the reconstruction of new identities

The application of biotechnology within the context of an industrialization and biochemicalization of agriculture reinforces a reorganization process in which traditional similarities and differences between persons, groups and sectors are disappearing, while also new identities are emerging. Although a complete picture of that reorganization process cannot be given here, three changes can shortly be mentioned, illustrating that an *emptying out* of traditional identities takes place and that *disjunctures* are created between the subjective and objective reconstruction of identities.

(1) *Farmers and the devaluation of their knowledge.* The industrialization of agriculture emerges not only because of the appropriation of some biological activities of farming by industries (Goodman, Sorj and Wilkinson 1987), but most of all by reorganizing and re-linking new activities to industries. A reorganization of, for example, soil fertility through chemical fertilizers or the reproduction of crops by the development of hybrid seeds has gradually changed the position of farmers; not only because *farmers became part of the agro-industrial production chain*, but also because the industrial appropriation of the controllable biological elements of farming implies *a re-division of the expertise and knowledge base of farming*. The historical, scientific process to disconnect crops from their environments implies that farmers – the original experts of the biological elements of farming – become less expert. Their knowledge about their (traditional) good farming practices is *devaluating* and their craftsmanship in, for example, the reproduction of crops is no longer indispensable for good farming (Van der Ploeg 1991). Farmers need increasingly to be informed how to deal with these 'new' crops and become increasingly dependent on the external inputs, supplied by life-science companies. The process of disconnecting crops from their environment not only implies economic changes, but also *the production of ignorance at farmers level, the creation of a new market for information and the entrance of a new group of actors* (the extension officers). This changed position of farmers within global food chains implies that a critical analysis is needed for the presently popular and even populist call for participatory plant breeding and bottom-up approaches. Reflexive activities are required to find out how, in the context of biochemicalization of agriculture, devaluation of farmers' knowledge and industrialization of farming activities, participatory technology development can really contribute to a

social-economic empowerment of farmers. The reorganization of food production might imply that the traditional meaning of participatory plant breeding is emptied out and that a new content of participatory approaches needs to be formulated.

(2) *Scientists and the commodification of fundamental research areas.* The development of specific scientific trajectories, orientated to disconnect agricultural crops from their environment, also has repercussion on science itself and on the positions of scientists working in the different domains of biotechnology. For instance, the linear line from fundamental to applied research is vanishing. *Fundamental research areas become increasingly commodified* and integrated in a reorganizing food chain. The collection, selection and isolation of genetic information to be used to construct various in-built plant protection systems become economically valued (applied) activities. Also the techniques of transferring this information into plants or micro-organisms become highly valued and this explains why small biotechnology companies (for example Plant Genetic Systems) with some patents on these basic transformation techniques are bought by multinationals for millions of dollars.

The commodification of the originally fundamental research areas illustrates the appearance of an *extensive seed production sub-chain* in which commodities like restriction and ligase enzymes, plasmids, vectors, DNA sequences and host cells are produced by fundamental and applied, public and private research institutes. These new commodities flow from fundamental to applied, from public to private institutes and vice versa, all contributing to a further biochemicalization of agriculture, emptying out the original divide between fundamental and applied research.

Not only the traditional differences between fundamental/basic and applied research is vanishing. The same holds for the difference between *public and private* research institutes. Stimulated by the industrial appropriation of farming activities, one can notice that public institutes often do research on issues that are prescribed by the logic of a further modernization process of agriculture. The divide between public and private research agendas are increasingly fading away, because public research institutes maybe tended to solve problems, for instance, in the crop rotation systems, in the same way private research institutes do. The pressure on public institutes to follow the logic of modern agriculture can become so big that, even when they have no research contracts with private industries, they may follow the same research trajectories as these companies.

It is no longer a differentiating element whether the research is funded by public or private money, at the moment, when all these different institutional forms of research become increasingly linked to the homogenizing development of separating agriculture from its environment. The hybridized use of fundamental/applied, public/private research activities within the reorganization process of global food chains does not only lead to an *emptying out of their traditional meanings and differences*, but may enable biotechnologists to give *a new meaning to their work*. Although in general the biotechnologists subjectively consider and describe themselves still as being applied or fundamental researchers, working in a public or private research institute, *objectively* we can individuate some modifications in their working situations that may influence a redefinition of their identities.

First, the decentralized and *multiple forms of cooperation* between various research institutions within an increased extensive seed production chain also mean that the researchers *lose sight of their specific positions in the reorganization of food production*. Through the complex social organization of their scientific research the biotechnologists are *alienated from the social significance of their work*; at the moment that their scientific work becomes socially, increasingly, important. Within this new, contradictionary situation the biotechnologists construct subjectively two opposite images of themselves. On one hand, they return to the old image of being 'free-floating intellectuals'; able to make free intellectual choices, only constrainted by some vaguely defined social influences, as the presence of model systems and the influence of patenting issues on research. It is a self-image of being a scientist working within the huge walls of the laboratory, able to construct freely their own research agenda. On the other hand, biotechnologists emphasize their role as being the social actor per se. As one Wageningen plant breeder researcher has expressed: 'We are the basis of everything in agriculture. We really decide what happens in agriculture.' It is a construction of identity referring to the huge ambition of becoming another Norman Borlaug, the famous plant breeder and Noble Peace Prize winner, because of its contribution to the Green Revolution. It is the image of being able to construct 'miracle seeds' or 'seeds for global development', and reflects the important role that the externalized scientific development is indeed playing. Despite the internal contradictions, both images, however, are also interrelated to each other. They illustrate the bipolar positions that either intrinsically scientific/technological or social factors determine a research trajectory. Both images represent a *deterministic perception* of a *general role* that biotechnologists might play, neglecting the possibility of an identity formation of biotechnologists in relation to *the specific products* that they are developing.

A *reflexive identity* of the biotechnologist may start with the question whether his/her *specific* biotechnology product contributes to political control (for example, through the development of hybrid seeds or terminator seeds) or reinforces social transformations (for example, through apomixis or tissue culture). Just at the moment that *specific* biotechnological developments play an important political-economic role with completely opposed political orientations (increase of power or equity), social categories are still formed on basis of general (pro-contra) evaluations of biotechnological developments. The presence of this discrepancy implies that the construction of a reflexive identity can no longer be formed in relation to this general, deterministic vision on biotechnology, but requires a critical empirical reflection on the *particular products* that are developed.

(3) *Companies and the politicization of production.* The cooperation between public and private research institutes within a new fluid sector of life-sciences institutes will increasingly lead to the design and use of a large diversity of plants, such as herbicide-, insect-, disease-tolerant, vitamin and mineral enforced and nitrogen fixating crops. Around these different crops an increasing number of differentiated agricultural production systems can be created. However, the development of different crops and production systems can be related to the same biotechnological innovations. Just like in the past agrochemical companies stimulated the regional production of various chemical formulations from their patented chemical ingredients (Perelli, Flematti and Ruivenkamp 1984), the life-science industries can nowadays apply patented plant breeding techniques (for example, the *Agrobacterium* transformation method) for making attuned varieties and crops (for example, herbicide-, insect- and drought-resistant crops) for different agricultural production systems. The implication of this new interrelationship between *basic* plant breeding techniques, patented by a small groups of companies, and application in *different* crops for many different regions, is that various identities are changing.

First, there is a *shift in the identity of the persons that carry out the political control* on and programming of the production system. Political control on agricultural production is no longer primarily executed by an external institute (for example, the Ministry of Agriculture) but becomes primarily incorporated *in the organization of the production*, that is, formed through the supply of new seeds and their built-in plant protection systems. In other words, the ongoing reorganization process expropriates the political control from the domain of policy-makers and directly re-links it to the social organization of production. To indicate this new content of political identity I have spoken about 'the politicization of production' (Ruivenkamp 1989).

This new system of political control, directly embedded in the social organization of production, also *changes the identity of major life-sciences enterprises* and institutes. By effecting changes in the genetic structure of a plant, these companies will increasingly be able to determine where, when and how a crop should be sown, harvested and processed. These companies become *real political actors* that programme agricultural production and are able to impose a specific farming system and regulate the social organization of the agricultural production process *from a distance.*

The possibility of applying the same patented basic plant-breeding techniques for the development of different crops in different regions (for example, herbicide-resistant crops in the South and high-yielding organic crops in the North) also modifies *the reproduction of this controlling system.* This new political system is no longer primarily reproduced by global extension of these new seeds. On the contrary, the political stability of the reorganized production system may increasingly become dependent on the *segmentation of the production in regionally differentiated production systems.*[2] However, this development towards regionally differentiated production systems may also become a new base on which reflexive activities may be formed.

The Separation of Agriculture and Food Production

Regional food security has not only been determined by the presence of high-yielding agricultural production systems, but also by the regional experiences of conserving and processing the agricultural produce. Each region developed specific methods to maintain its agricultural produce (for example milk) which led to many regionally different forms of food products (for example, various types of cheese). Global food production has long been characterized by these two contrasting developments: an increasing regional *uniformity of agricultural production* by cultivating crops like wheat, corn, rice and so on and a regional *diversity of techniques* to conserve and process the agricultural produce in food products (Perelli, Flematti and Ruivenkamp 1984).

The regional diversity of techniques was originally primarily applied on-farm and became gradually appropriated by small enterprises in the region. Through the up-scaling and scientification of locally specific conservation techniques and through the technical developments in different processing branches, the enterprises could manufacture an ever growing diversity of food products in each region. However, the industrialization of food processing also implied a shift in the processing techniques, *from conserving towards transforming the agricultural produce*. This gradual, *quantitative,* shift in the processing techniques – increasing while the industrialization of food processing was increasing – also changed the agricultural sector, the processing industry, their interaction and their identities *qualitatively.*

The agricultural sector was gradually reorganizing itself in such a way that it was brought into harmony with the new demands of the industrial process of conserving and manufacturing the agricultural produce. As a consequence the agricultural sector became increasingly a *sub-sector of the food processing* industry that could conserve their products longer and organize themselves increasingly at an *international level*. They became great international players and re-adapt their processing techniques to the new international context in which they were operating. This re-adaptation of processing techniques was developed for example by Unilever, one of the first internationally operating food companies, producing the same end-product (butter(ine) or margarine) by using other agricultural raw materials (no longer milk fat, but plant oils). No longer the conservation and processing of the agricultural produce but *the extraction and manipulation of the food components* from a broad spectrum of agricultural products became the core of a new relationship between agriculture and food processing.

While the growth of national food companies was related to the enlargement of regionally developed processing techniques, the growth of the international food companies is increasingly based on the *disconnection of the food products from its agricultural produce*. This disconnection is realized by another quantitatively change in the techniques of processing: *from transforming towards splitting up the agricultural produce and re-aggregating* the food components into traditional food products – a modification in the processing techniques that takes place through the increased use of biocatalysts that are able to extract the basic food nutrients from a broad spectrum of different agricultural sources. This shift from transforming into splitting up the agricultural produce in food components made it also possible to consider agricultural products as *interchangeable sources of raw materials* of which several packages of food components can be assembled at global level. Biotechnology reinforces quantitatively the disconnection of the agricultural and food products and transforms it qualitatively into a *separation* of food from the agricultural produce through:

- the development of new enzymes able to extract basic nutrients like carbohydrates, proteins and fats from an increased spectrum of different crops;
- the development of microbiological processes to produce these components by microorganisms in the factory.

These two developments reinforce the ongoing reorganization process of separating food products from its specific agricultural produce, but may also create a new social space for reflexive actions with respect to these developments.

Interchangeability of products and new identities of actors and actants
A complete picture of the reorganization process in the third phase of the food chain cannot be given here (see Ruivenkamp 1989). We limit ourselves to describing some changes that clarify how the traditional meanings of various categories have been emptied out, and how this reorganization process at the same time creates space for various actors to start reflexive actions and re-aggregate themselves around new categories.

(1) *The food product: from the processing of a specific agricultural product towards an assemblage of industrial food components.* The disconnection of food and agricultural products has first of all changed *the identity of food products* qualitatively. The new production system of food tends to manufacture a number of nutrients like fats, proteins and carbohydrates that are assembled from a broad range of agricultural sources. Additives like flavouring and colouring agents are added and this packet of ingredients is put together in a recognizable form of a *traditional food stuff* and sold to the consumers. One implication of this restructuring process is that agricultural products are increasingly conceived as a totality of biochemical components that may be used as general inputs for the food processing industries. Another implication is that the food products, increasingly separated from their specific agricultural sources, can increasingly be re-linked to other signifiers and be placed in the economies of signs (Lash and Urry 1994).

(2) *The producers: emptying out their differences.* The collapse of the linear and regional integration of agriculture and food processing, in which a specific agricultural product is processed into a specific end-product, also makes an end to the *transparency* of the food production. The 'liberation' of the food manufacturing companies from a specific supply of agricultural *products empty out the differences between crops and their producers.* Through the industrial application of enzymes, it does not matter anymore whether a sweetener, such as glucose, is made from sugarcane, corn, or

sorghum. All these plant materials can be used as a glucose source for the production of, for example, beverages. Producing food products by interchangeable agricultural sources implies that the borders between different crops and agricultural sectors fade away. Not only food production becomes non-transparent, but also political changes take place. The traditional organization of farmers and workers around specific food crops (for example, the sugarcane plantation workers) lose their political significance, at the moment that biotechnological developments make it possible to substitute sugarcane by high-fructose syrup, made from corn, or by aspartame, produced by micro-organisms (Ruivenkamp 1986). The reorganization process leads to a *political involution of those organizations* (farmers organizations and unions) that still aim to aggregate their members around these (vanishing) sectors. Even the concept of agriculture as a sector – as a reference point for identities and political organizations – enters in crisis at the moment that industrial products from micro-organisms become competing products for the products cultivated by farmers. The objective reconstruction of, for example, the sugar sector into flexible production circuits of sweeteners makes a continuing subjective construction of identities at horizontal level, such as on crop level (for example, sugarcane farmers), problematic. Although the farmers often perceive themselves still as sugarcane or corn farmers, the flexible use of agricultural raw materials and biochemical components by food companies urges farmers' organization and food unions to reflect on which basis new collective political identities and organizations may arise.

(3) *Biochemical industries as food companies*. The blurring of traditional differences among agricultural products (for instance, between sugar and corn as sweetener source) liberates farmers from their social categories of being, for example, corn farmers and replaces these farmers in new fluid entities. Besides, the power relations between farmers organizations and food enterprises within the international food system are modifying (see 2). Finally, the separation of specific food products from the agricultural produce implies *the entrance of the biochemical industries in the food chain*. While the food manufacturers will increasingly be based on the supply and use of interchangeable food components for the manufacturing of their food products, they also become increasingly *dependent* on the biochemical industries that deliver those catalysts to extract the food components from various agricultural raw materials. The biochemical industry, which supplies these biocatalysts, will become an actor in *the restructuring of the food production*. The manufacturing of these biocatalysts, like enzymes, yeasts and so on, becomes a crucial *political activity*. These products themselves become actors (actants) or as I

called them '*politicizing products*', able to influence or even to programme the organization of various food production systems.

The companies developing these catalysts are increasingly getting a *political identity*. They have at their disposal the scientific and technical capacities to make a growing number of farm crops and industrially manufactured food components mutually interchangeable. They have in their hands the mechanisms to divide and control the producers of these basic nutrients. They can decide which and how an increasing number of farmers, workers and micro-organisms in the different sectors in different countries can be played off against each other. Through the separation of food products from its agricultural produce, these catalysts obtain an increasing 'politicizing' influence in the reorganization of food production. However, this dual process of disappearing traditional social categories together with the emergence of 'politicizing products' may stimulate actors to form new groups, networks or organizations in which they redefine their identities, especially in relation to a reconstruction of these 'politicizing products'.

(4) *Redefinition of locality*. New enzyme applications and microbiological production of food components may increase the possibilities of manufacturing various food components from locally available agricultural and industrial products. Dependent on the application of the scientific advances and of specific price calculations, food production may increasingly be organized on a more regional level, using various techniques to process local sources into the desired combinations of food components. Each locality may increasingly be enabled to aggregate its own specific composition from various global/local flows of food components. An apparently increased *autonomy* of regional food supply at *local level* may arise. However, this will be related to an increased *dependency* on the use of globally developed biocatalysts. The increased production at regional level through the use of scientifically developed biocatalyst takes places within food conglomerates, assembling the interchangeable food components from agricultural and biochemical sources at international level. The international competition between these food agglomerates will no longer be a simply *competition* between for example different vegetable sources (products) but between their *regional methods* of producing and converting food components. The locally specific way of producing and assembling food components will become crucial for the competitive position of these globally operating food companies. They will consider the regionally different social forms of producing and converting food components as being *interchangeable production units*. This development will also gradually change the

traditional meaning of self-sufficient food regions. This can no longer exclusively be defined at production level. On the contrary, the relation with the globally developed biocatalysts should also be taken into consideration. This implies that the identity of locality can no longer exclusively be defined in geographically terms in relation to the local ways of producing food, but will increasingly be formed in relation to these new global/local flows of enzymes, biocatalyst and other biotechnologies. Therefore, the following section describes how communities aim to re-construct their local identities by re-appropriating various biotechnological developments.

THE DEVELOPMENT OF TAILOR-MADE BIOTECHNOLOGIES

The transformative effects of agro-industrial biotechnological developments to the effect that agriculture as a sector is vanishing, crop products are reduced to a sum of biochemical components, hybrid forms of fundamental/applied, public/private research are emerging, politicizing products are produced and patented, and companies become political agents, cannot but impact on the subjective identity construction of all actors involved in these new developments. In different regions 'villagers' (Magnaghi 2000) are trying to find a way out from the schizophrenic situation of these agro-industrial biotechnological developments. They feel no longer comfortable with the dualistic approach of either protecting themselves against the biotechnological innovations by refusing them completely or to accept by participating in these developments without changing them. Instead of this pro-contra position, characterized by a *lack of imagination* to change, some villagers in various regions follow a more constructive approach. They aim to develop a balanced and dialectical approach of integrating and indigenizing some (biotechnological) innovations, attuned to the local resources. The central question for such a balanced and dialectical approach is whether and how biotechnology as an exogenous instrument can be re-appropriated by local initiatives and become a catalyst for endogenous developments (Ruivenkamp 2001). Below, we describe shortly some concrete experiences in different countries to re-appropriate and attune biotechnologies to endogenous developments.

The Re-appropriation of Knowledge and Craftsmanship

In Andhra Pradesh, India, castor is one of the most important dry-land cash crops, burdened with a wide range of diseases and pests. One of the most damaging diseases is the presence of wilt, which is a fungal disease (Buijs 2002, p. 30) that needs large amounts of chemicals to be controlled. This approach is economically and environmentally unsustainable for the resource-poor farmers that grow castor. The Biotechnology Unit (BTU) of the Institute of Public Enterprise of the Osmania University Campus, Hyderabad, together with research institutes and farmer groups in Nalgonda and Mahaboobnagar districts, are elaborating an initiative to protect the crop against these plaques by the development of *bio-pesticides, made from locally available resources*, such as local species and strains of a fungal family, called *Trichoderma* and *directly produced by the farmers*. Characteristic of this initiative is that the knowledge and craftsmanship of farmers have not been denied but recognized and re-valued. Farmers have been consulted to find and select the specific native strains of *Trichoderma* as a biocontrol agent to protect castor against wilt-disease in this region. And because the *selected Trichoderma strain is specific to the target region*, and thus specific to the local strains of the wilt causing *Fusarium* fungus, it increases its regional efficiency. It makes it more wanted than the commercial products because these products contain universal strains with a less specific virulence to the local strains of wilt *Fusarium* than the locally selected and developed strains. The project, however, is not only characterized by the *revalidation of farmer's knowledge in selecting regionally specific pathogens for the crops,* but also by a *reorganization of the production of bio-control agents.* The bio-control agents are produced *directly at village level* in which increasingly cheap local resources for its production are used. Another advantage of the farmers' selection of these specific *Trichoderma* strains is that these strains are fast-growing fungi that can be grown in many different substrates. This offers the opportunity that local resources, like several carbohydrate sources and waste products, can be used as a substrate. In other words: the villagers try to reverse several aspects of the biochemicalization process; there is no longer a devaluation but revalidation of farmers' knowledge about the soil–crop interactions. Through the integration of their knowledge in the selection of regional specific pathogens, they also become able to control and organize the production of these bio-control agents. Through these processes a subjective reconstruction of identities takes place changing over from an identification with traditional professional and sectoral categories like farmers, producers and consumers towards a more holistic category like 'villagers'.

The De-commodification of the Seed Production

It has been argued that the reorganization of global food chains goes together with an extension of the seed production subchain and that within this subchain several (fundamental-applied) research areas are increasingly commodified. Through the industrial appropriation of these areas farmers become increasingly dependent on life-science industries, that are able to carry out a remote control on agricultural production through the supply of specific seeds, such as hybrid seeds or seeds with specific built-in protection mechanisms.

One new biotechnology application that might reduce these farmers' dependence on hybrids is apomixis: the production of seeds without fertilization. Many wild plants are naturally apomixtic (for example, the common dandelion, *Taraxacum* sp., but not the domesticated crop species). The introduction of this trait to crop plants would allow them to produce seeds with identical genetic properties, fixing the plants genetic make-up. Apomixis could be applied to obtain seeds from a hybrid plant, while retaining the vigour of the first generation. Apomixis could provide small farmers with a tool for removing the hybrid's biological protection against propagation. It could thus undermine the profitability of the private seed industry and permit many more of the world's farmers to use hybrids. The introduction of this trait into crop plants would restore farmers to their traditional role as innovators (Jefferson 1994, cited in Van der Sande, Ruivenkamp and Malo 1996, p. 185).

Research on *apomixis* in crops such as pearl millet, rice, maize and forage grasses has been carried out at various international agricultural research centres (IARC's) (Van der Sande, Ruivenkamp and Malo 1996, p. 185). However, concrete results are not known, yet. Of course, this research is a politicizing issue. On the one hand, companies may be interested to transform the hybrid's biological protection against propagation into a more general protection of the propagation of crops, either by applying patents on all the important characteristics of propagation, including the apomixis techniques, or to reinforce the biological protection against propagation (for example, terminator seeds). On the other hand, groups of scientists of various disciplines together with farmers' organizations may be interested to find concrete ways how the propagation of crops can come again under control of the farmers. Apomixis might be one *technological solution* to realize this objective.

Another strategy, less spectacular but maybe more convincing, is the attempt of *changing socially the production chain* of various reproduction techniques. Jantina de Vries (2002) describes in her thesis this possibility. She argues that the production chain of tissue culture plantlets in Ghana is in the middle of an ideological battle of two contrasting identity constructions of the Ghanaian scientists. One group of scientists refers to an ideal-typical tissue culture development in the North, trying to imitate that development in the South, despite all kinds of regional constraints as, for instance, the absence of a continuous electricity power supply. Another group of scientists aims to indigenize the tissue culture production chain and attune the productive organization to the objective of supporting locally specific endogenous developments. Because at this moment the Northern identity construction is still the most powerful one, de Vries recommends that Ghanaian researchers be enabled to learn from their Indian colleagues, where already successful attempts have been realized to de-commodify several phases of the production chain of tissue culture plantlets as, for instance, by substituting agar – delivered by industries – by other nutrients made of regional waste products processed by worms and manufactured by the farmers themselves at village level (Buijs 2002, p. 34).

Hybridization and Indigenizing of Biotechnological Developments

he locally specific attempts to indigenize biotechnological developments can only be described on basis of a specific analysis of various biotechnological developments. It is known that, for instance in India, the science-based tissue culture applications on local tree species as amla, tamarind, custrad apple and karaya, has been combined with people-based experiences of *applying the grafting technique* (Buijs 2002, p. 80). Besides, Indian NGOs made an important innovation in the propagation of neem and teak trees. Instead of going through the whole tissue culture phase, they cut auxiliary shoots, taken from the tissue cultured teak and neem tree clones, and put these shoots directly in vermiculite (a vermiculture product) and let the shoots root directly (Buijs 2002, p. 48). This medium is very cheap and the use of expensive agar-based medium can be avoided. The newly developed method is a hybrid between modern tissue culture technology and local vegetative propagation techniques (Buijs 2002, p. 48). Another example is the farmers' selection of local strains of *Bacillus Thuriengiensis* (*Bt*) and *Rhizobium* for their plant protective and growth stimulating characteristics. The Indian farmers also isolate and ferment the *Bt* material in field laboratories and distribute 'their' end-products in the form of pellets to other farmers in the village. It is said that the selected strains have better results than the commercial *Bt* products delivered by industries because of the very strong regional orientation of the farmers' selection procedure. This case shows that through the integrated activities of selecting, isolating, fermenting and distributing locally specific *Bt* material these farmers are re-grouping themselves and reconstruct their identities, going beyond the traditional/modern meaning of being farmers dependent on external suppliers.

Instead of referring to other examples, I prefer to emphasize that the hybridization of techniques also assumes the presence of specific *institutional re-arrangements.* Important lessons can be learnt from the experiences in India, where the Indian Council of Agricultural Research (ICAR), being aware of the distance between research institutes and farmers' practices, in the 1970s decided to establish Farm Science Centres or socalled Krishi Vigyan Kendra (KvK). The first KvK was established in 1974 and by 1994 261 KvK were functioning and it was decided to establish one KvK in each of the 500 districts of the country (Shukla, Reddy and Sanghi 1994, p. 9). The objective of these centres is to facilitate the process of technology transfer by providing *feed forward as well as feed backward information to technology development and research* (Shukla, Reddy and Sanghi 1994, p. 10). The Krishi Vigyan Kendras are the grassroot-level technology transfer and vocational training institutes, designed for bridging the gap between the available technologies at one end and their application for increased production at the other, the objective being to have an appropriate technology that may be economically profitable, ecologically sustainable, technically feasible and culturally compatible. This, according to one of the initiators of the KvK institutes, has led many times to a *mixture* of indigenous-technology-knowledge of local farmers with the package of technology, evolved by scientists (Shukla, Reddy and Sanghi 1994, p. 1). This blending of technologies continues as described in above-mentioned examples. Through the locally specific modification of global biotechnological developments not only farmers regroup themselves, but also new institutes appear in which also scientists construct a new identity. For example, the director of the KvK in the Nagonda district, having finished his PhD at an American university and having worked for a long period at Indian universities, decided to go back to his village and to work with other scientists and other villagers for the social and economic empowerment of the village. He considered himself no longer a beta scientist but primarily a social worker for his village, integrating technical and social dimensions in new tailor-made biotechnologies.

Going Beyond the Disconnections/Separations Embodied in Actual Agro-biotechnological Developments

For the transformation of biotechnology, as an exogenous instrument into a catalyst for endogenous developments, it will be crucial to find out how these village initiatives aim 'to go beyond' the separations that are embodied in the actual agro-industrial biotechnological developments. To end this section we will limit ourselves to a short description of initiatives in which it is aimed to relink agriculture to its environment and food products to the agricultural produce.

Linking Agriculture and Environment

The separation of agriculture from its environment, reinforced by plant biotechnological developments, is an ongoing historical process, but not an inevitable one and, therefore, not necessarily leading to an increased biochemicalization of agriculture. On the contrary, the biochemicalization of agriculture is increasingly criticized in view of its negative results at economic and environmental level. Various reflexive activities at village level are developing, trying to create space for pluriform approaches to reinstall a new relationship *between the changed agricultural production systems and the changed environment.* For example, 'new varieties' are developed through which also 'other changes' in the relations between agriculture and environment are catalysed. In the design of these 'new varieties' several social issues are directly taken into consideration, such as the possibility of changing the crop rotation system and to reduce the ecological problems of actual agricultural production systems and to increase the income of marginalized farmer groups. These new varieties are illustrative for *a social/technical ensemble approach, in which the variety itself catalyses certain social changes.* A social/technical ensemble approach that differs radically from the traditional mono-crop design approach. One concrete example of such a 'new illustrative variety' with its catalysing influence on endogenous developments is *the dual purpose, early maturity sorghum variety* (*CSV 15*), developed in India by the National Research Centre for Sorghum (NRCS) of the Indian Council of Agricultural Research.

Sorghum is one of the most important cereal crop for poor people and is grown for food, feed and industrial products. The development of a dual purpose variety gives these farmers the opportunity to improve both the food and feed productivity. Still more innovative is that it also concerns an early maturity crop. Instead of 144 days of the recently released varieties, this new variety can be harvested after only 112–115 days (Rana 2000), which has enabled the farmers to plant chickpea almost one month earlier as a rotation crop, minimizing the incidence of wilt in chickpea and reducing the use of chemicals. The short duration, dual purpose sorghum variety enables farmers to get an assured higher grain yield; to get a higher monetary return because of its dual (food/feed) purposes and ensures more income because of the improved rotation with chickpea. Instead of harvesting one unsecured, long duration local sorghum variety, farmers can now cultivate and get income from two assured crops in a year. The hybridity of technical and social dimensions incorporated in this new 'illustrative variety' may reinforce other reflexive activities of scientists and farmers groups. It may stimulate them to develop new laboratory products, in which the reinforcement of locally specific endogenous developments is directly incorporated in its social/technical characteristics. It goes without saying that these new varieties can also become catalysts for reconstructing identities. In particular, beta scientists may regroup themselves and develop new ways of placing social issues at the core of their research.

Linking Food Products and the Agricultural Produce

Along the historical development of making agricultural products interchangeable raw materials for food processing industries, extended food production chains emerged at international level. At regional level the direct line between food and agricultural product is vanishing and food products are increasingly a sum of biochemical components (such as carbohydrates, proteins, fats and oils), functional components (such as vitamins and minerals), flavours and additives. In reaction to the new organization of food production, reflexive activities have emerged. In the North several NGOs ask for more transparency in the way the products are produced and ask for more safety controls on the quality of the industrialized food products. Also in the South and North new regional initiatives have emerged to re-link the locally specific food products to the locally specific agricultural produce. This latter implies a specific redesign of biocatalysts (enzymes and starter cultures) in such a way that more food can be produced and consumed locally. One example is the development of a new starter culture by the Food Research Institute in Ghana, which makes it possible to accelerate the fermentation of maize dough in two locally important products, kenkey and banku (de Vries, 2002). Another example is the production of fungal spawn for mushroom production at village level (de Vries 2002). By developing these locally specific biocatalysts, in which global and local technological developments are mixed, new direct lines between food and agricultural production and food production and consumption are restored at local level. In general most biocatalysts are still orientated to disconnect food and agricultural production. A transformation of that script in the actual biocatalysts into a script orientated to empower the local potentialities of locally specific food productions may enhance the reconstruction of identities of scientists, consumers, producers, farmers and others.

CONCLUSIONS

The aim of this chapter was to conceptualize biotechnology from a dialectical perspective, showing that biotechnology contains homogenizing as well as heterogenizing elements and that through these contrasting influences not only the (social) landscape of agriculture and food production is changing, but also the personal contexts of various actors, such as farmers, scientists, and companies. It has been argued that there is an increased de-synchronization between the subjective and the objective reconstruction of identities and that through this discrepancy new forms of political control are emerging. New mechanisms to programme production are no longer designed in separated institutes, such as ministries, but are directly embedded in a specific socially organized production system, in which the use of new agricultural inputs, enzymes and microbiological processes play a crucial role.

Biotechnology is socially such an important technology because it develops these new politicizing products for the reorganization of food production and because the development of these politicizing products takes place within a very peculiar context. It has been argued that the politicizing products are developed within a complex social organization of scientific research in which new hybrid forms of fundamental/applied, public/private research institutes are emerging. This dual development of a complex organization of scientific research and the emergence of politicizing products in global food chains creates the contradiction that the biotechnologists are alienated from the social significance of their work at the moment that their work becomes socially increasingly important. It illustrates the phenomenon that the assumed duality of social developments and actors' sense-making and reacting has been undermined by the specific form of the reorganized food production chain. However, the dialectical perspective urges us to look for a new space – within the ongoing reorganization process – for criticial reflexivity. It has been emphasized that a critical reflexivity can no longer be formed in relation to a general deterministic vision on biotechnology, but requires a criticial empirical reflection on the *particular products* that are developed. The duality of social developments and actors' sense-making has to be re-found in the development of these particular, politicizing products.

The second section has given some examples of initiatives that tailor particular biotechnological development by mixing different beta gamma knowledge systems and incorporating institutional re-arrangements in their initiatives. From these particular experiences it can be learned that a general strategy is followed of 'going beyond' the separation processes that are actually embodied in agro-industrial biotechnological developments. Tailor-made biotechnological developments in various regions are characterized by their attempts to relate in its technical dimensions the social aspects of re-linking agriculture and environment, re-linking food products and the agricultural produce and the agricultural product with its food quality. This ongoing process of disembedding tailor-made biotechnological developments from the separation processes and re-embedding them within locally specific developments also implies a reconstruction of actors identities. We have shown that in this process many traditional social categories disappear. Farmers become chemical producers and fermentators of *Bt* strains, while beta scientists became social workers for villages. However, we have also emphasized that in general the actors still view themselves primarily as being farmers, beta or gamma scientists, working in public or private institutes, neglecting all the hybrid forms that play a crucial role in the reorganization process of food production. Precisely, because of this discrepancy between the objective and subjective constructions of identity in the actual organization of food production, it becomes crucial to search for new spaces in which new identities can be formed. Often a lack of imagination prevents the envisaging of these new possibilities and stimulates a 'go back' to the old categories. We hope that sufficient arguments have been given to question such an (old) identity description and that it has become clear why a reconstruction of identities may take place around the development of particular biotechnological products. Therefore, an international programme on tailor-made biotechnologies has been set up with the objective of exchanging information and facilitating the emergence of new global–local combinations of social/biotechnological ensembles, as well as facilitating the emergence of new identities at glocal level around the new politicizing products. This programme tackles the new challenge to find ways of transforming the agro-industrial biotechnology as an exogenous instrument into a catalyst for endogenous developments and to stimulate the development of new (hybrid) identities at glocal level within the new fluid networks of particular biotechnological developments.

NOTES

1. I want to thank Peter Leisink, Henk Overbeek, Alphons Rommelse, Ab van Eldijk and Jacob van Etten for their comments on the first drafts of this chapter.
2. Therefore, reflexive action is also needed on the actual critical content of the request of some consumer and environmental NGOs to ask for a segmentation of food chains, like GMO and GMO-free food chains, within this context of a biochemicalization of agriculture.

REFERENCES

Appadurai, A. (1990), 'Disjuncture and difference in the global cultural economy', in M. Featherstone (ed.), *Global Culture: Nationalism, Globalization and Modernity*, London, Sage, pp. 295–310.

Beck, U., A. Giddens and S. Lash (1994), *Reflexive Modernization: Politics, Tradition and Aesthetics in the Modern Social Order*, Cambridge: Polity Press.

Belforte, S., E. Merli, M. Vogliazzo and A. Zeppetella (1976), 'Dalla Citta – Fabbrica alla Fabbrica Diffusa', in *La Fabbrica nella Societa. Quaderni del Territorio*, Milano: Celuc Libri.

Bull, A. T., G. Holt and M. D. Lilly (1982), *Biotechnology, International Trends and Perspectives*, Paris: Organization for Economic Cooperation and Development (OECD).

Buijs, J. (2002), 'Shaping biotechnological developments in the context of resource-poor farming: a study of the Andhra Pradesh–Netherlands Biotechnology Programme in Andhra Pradesh, India', MSc thesis, Wageningen.

Crouch, C. and A. Pizzorno (1977), *Conflitti in Europa, lotte di classe, sindacati e stato dopo il '68*, Milano: Etas Libri.

De Vries, J. (2002), 'Biotechnology in Ghana: an exploration of the impact of the international ruling dogma's on the direction of development of biotechnology in a developing country and the perspectives for change', MSc thesis, Wageningen.

Goodman, D., B. Sorj and J. Wilkinson (1987), *From Farming to Biotechnology: A Theory of Agro-Industrial Development*, Oxford: Blackwell.

Hannerz, U. (1992), *Cultural Complexity: Studies in the Social Organization of Meaning*, New York: Colombia University Press.

James, C. (2001), *Global Review of Commercialized Transgenic Crops: 2001*, New York: Ithaca, ISAAA Briefs, no 24.

Lash, S. and J. Urry (1994), *Economies of Signs and Space*, London: Sage.

Magnaghi, A. (2000), *Il progetto locale*, Torino: Bollati Boringhieri editore s.r.l.

Negri, A. (1979), *Dall'operaio massa all' operaio sociale*, Milano: Multhipla Edizioni.

Perelli, A., F. Flematti and G. Ruivenkamp (1984), 'L'agribusiness multinazionale', *La Questione Agraria*, **15**, 53–76.

Pistorius, R. and J. van Wijk (1999), *The Exploitation of Plant Genetic Information: Political Strategies in Crop Development*, Amsterdam: Print Partners Ipskamp.

Rana, B. S. (2000), *Annual Report National Research Centre for Sorghum*, Rajendranagar, Hyderabad: Indian Council of Agricultural Research.

Robertson, R. (1995), 'Glocalisation: time–space and homogeneity–heterogeneity', in M. Featherstone, S. Lash and R. Robertson (eds), *Global Modernities*, London: Sage, pp. 25–41.

Ruivenkamp, G. (1986), 'The impact of biotechnology on international development: competition between sugars and sweeteners', *Vierteljahresberichte des Forschungsinstituts der Friedrich Ebert Stiftung*, no. 103, March, special issue on 'New technologies and Third World development'.

Ruivenkamp, G. (1989), *De invoering van biotechnologie in de agro-industriele productieketen. De overgang naar een nieuwe arbeidsorganisatie*, Utrecht: Jan van Arkel Uitgeverij.

Ruivenkamp, G. (2001), 'Access to food through tailor-made biotechnologies', Wageningen, TAO document.

Shukla, A. N., G. G. Reddy and N. K. Sanghi (1994), *Intervention of Krishi Vigyan Kendra (Farm Science Center) in Transfer of Technology: A Case Study of K.V.K., Gaddipalli*, Nalgonda, New Delhi: Indian Council of Agricultural Research (ICAR).

Van de Sande, T., G. Ruivenkamp and S. Malo (1996), 'The socio-political context', in J. Bunders, B. Haverkort and W. Hiemstra (eds), *Biotechnology: Building on Farmers' Knowledge*, London: Macmillan Education, pp. 181–98.

Van der Ploeg, J. D. (1991), *Landbouw als mensenwerk: Arbeid en technologie in de agrarische ontwikkeling*, Muiderberg: Dick Coutinho B.V.

Van der Ploeg, J. D. and A. Long (eds) (1994), *Born from Within: Practice and Perspectives of Endogenous Rural Development*, Assen: Van Gorcum.

Vergopoulos, K. (1986), *L' impact des nouvelles technologies sur les industries alimentaire europeennes*, Brussels, Commission of the European Communities, Directorate General for Science Research and Development, December (Fast Occasional paper no. 126).

Yoxen, E. (1983), *The Gene Business: Who Should Control Biotechnology?*, London: Pan Books.

14. Organizational Relationships in the Networking Age: A Concluding Analysis

Willem Koot, Peter Leisink, Paul Verweel and Paul Jeffcutt

INTRODUCTION

As a framework for presenting and discussing the main findings of this volume we will begin by summarizing the central theses and questions which we took from post-modernity theories. We will then present the main findings, discuss their meaning in the light of these theories and reflect on what they say about the empirical and intellectual foundations of management and organizational knowledge. The final section will elaborate directions for future research. It argues, first, that it would be worthwhile for further research to draw on cognitive anthropology and conceptualize culture and identity as human capacities to reframe cultural knowledge and to survive in a shifting cultural force-field. It also argues the importance of comparative research on issues of identity formation and organizational bonding in the networking age in non-Western countries, in order to learn whether the ways in which, for instance, black Africans cope with cultural complexity are different from the ways Europeans and Americans do. Finally the relevance of transdisciplinary knowledge production is reflected on, both in the light of the continuing process of unbounding and pluralization of the socio-economic order and the everyday world of complex organizations, and in the light of the further unbounding and pluralization of the human sciences.

NETWORKING, ORGANIZATIONAL COMMUNITIES AND THE DYNAMICS OF IDENTITY FORMATION

Chapter 1 drew on theories of late- and post-modernism to frame the central questions of this volume. Following authors such as Bauman, Giddens, Hall and Sennett we adopted as central supposition that social transformations at the end of the 20th century are dislocating modernity's traditions, institutions and identities, and are thus undermining the frameworks which gave individuals stable anchorage in the social world. Our interest in the ways in which work organizations and organizational actors would (re)act to these changes means that we have a more specific focus than post-modern theories in general. The latter are interested in what are considered epochal changes, such as the collapse of communism, the decline of the nation-state in relation to globalization, the transformation from mass consumption patterns to lifestyle niches and the pervasiveness of (mass) media, and in the impact of these changes, which is supposed to range from the fragmentation of social classes and the loss of social cohesion to social forms and identites marked by fragmentation, multiplicity, plurality and indeterminacy (Hall 1992; Thompson 1992). Their interest is thus in social and cultural changes in post-modern societies in general, whereas we are interested in the social and cultural consequences of changes in the economy and work organizations referred to by 'the networking age'.

Our focus means that we share the interest of researchers studying the effects of post-industrialist or post-Fordist transformations of economy and work organizations but rather than focusing on the techno-organizational changes – involving, for instance, outsourcing, delayering organizational structures, networking within and between organizations – we are interested in the social aspects, especially the dynamics of identity formation and bonding which emerge in interaction with these transformations. For example, one may suppose that the Fordist organization of the mid-20th century which combined mass production with mass worker social policies and mass trade union organization, represented a symbolic community which was able to generate a strong sense of identity and allegiance – or rather dual identities and allegiances – and that the breaking up of this bureaucratic model and its replacement by a fragmented, more flexible networking model resulted in a loss of identification with the organization and a crisis of identity. Thus our concerns are related to the main stream of post-modernist writing because we are interested whether the consequences which are expected by this body of literature in terms of fragmentation, loss of social cohesion and social forms and multiple identities, can also be observed in the organizational relationships which we study.

In fact, two different sets of hypotheses can be elaborated. First, one can suppose that the fragmentation, plurality and indeterminacy caused by social transformations in general and by post-Fordist transformations of economy and work organizations in particular result in a crisis of identity on the part of managers, employees, suppliers and other stakeholders, and that their identification with the organization with which they have some sort of contractual relationship (as employee, supplier or whatever) is eroded. This set of hypotheses is broadly in line with Sennett (1998) who argues that the new flexible capitalism prevents employees from entering into any lasting forms of organizational relationships, from feeling at home and even from developing a coherent life story. Second, one can also suppose that because social transformations result in a fragmentation of social classes and loss of social cohesion, organizations are in a position to become a community with which managers, employees, clients and others can identify. On the one hand, organizational changes introducing all sorts of flexibility act against organizations becoming such a community but, on the other hand, there are organizational changes such as teamworking, empowerment, organizational participation and deep leadership which can offer a basis for meaningful work and meaningful relationships with colleagues and line managers (cf. Grint 1997; Heller et al. 1998). Indeed, the past decade has seen an increase of research and management interest in commitment and identification, in bonding employees, in new psychological contracts, in trust within and between organizations (see, for instance, Lane and Bachmann 2000; Meyer and Allen 1997; Rousseau 1995). This management interest has been understood as supporting this second set of theses, while protagonists of the first set of hypotheses regard this management interest as pure rhetoric and no more than an attempt to compensate for the negative effects of organizational transformations. This summary of questions and theses, which are drawn from post-modernist theories and current research of organizational relationships, serves as a framework for presenting and discussing the main findings of the contributions to this volume.

These contributions do not generally confirm the first thesis which suggests a crisis of identity and an erosion of identification with the organization, but neither do they support the idea of organizations as communities with which managers, employees, clients and others can identify and can have meaningful relationships. Our findings are more complex and demonstrate continuities as well as discontinuities.

In the case of organizations where turbulence and fragmentation are most apparent – as in the former public telecom bureaucracies, which are studied by van Marrewijk – or in the case of organizations which are regarded as virtual themselves – such as the 'Dotcoms' studied by Veenswijk – processes of identity construction and identification appear by and large to pursue familiar lines. Van Marrewijk shows the relevance of the concept of nested identities, developed by Ashforth and Johnson (2001) who claim that lower-order identities (job, work group, department) tend to be subjectively more salient than higher-order identities (division, organization). Indeed, the focus of identification of telecom employees continues to be their job, work group and department, although what and who this involves changes because of the ongoing organizational restructuring processes, which at some stage turned full-service customer-orientated regional units into central technology-based business units. Yet this process of splitting up the former bureaucratic organization into a network of business units and the simultaneous process of entering into a network of international alliances seems to have gone along with a decline or loss of identification of telecom employees with the company and the top management. Supposedly the effort that telecom employees had to spend in (re)negotiating their identities in relation to their 'new' colleagues, in combination with rivalries between business units and with what 'traditional' employees perceived as the disruptions and intransparence of the strategies of those at the top, made them concentrate on their micro-world and withdraw from the macro-organization. Similarly the case studies of 'Dotcom' organizations done by Veenswijk give only limited support to the assumption that virtualized work connections lead to the formation of hyper-real marketplaces of semiotic exchange and to a pluralization of subjectivity. In fact, identity construction in 'Dotcoms' appears to focus on groups of professionals and is situational in the sense that the incentives to connect and identify with specific subgroups in the organization are closely related to the state of external connections (for instance, with clients and banks) at the level of the organization as a whole.

Other contributions add to the overall impression that changes in organizational practices take place and disrupt familiar routines to more or less extent, but also that managers and employees adapt to those changes and make sense of these changes and themselves without ending in a crisis of identity. Thus Kamsteeg and Koot, in their analysis of managers, show that management still aims at control but that the means have shifted to managing the meaning of business processes for employees. Although this shift implies that the tasks required of the manager are making greater challenges upon him or her, in terms of managing others as well as managing themselves, managers survive, albeit by drawing on familiar utilitarian forms of control rather than on reflection. Keenoy analyses the impact of the introduction of managerial control in British universities, which he regards as most alien to such control regimes because of the traditional value of academic freedom, and is interested in how academics (re)construct their new social reality. He finds that they do not appear to operate on the basis of fixed ideal-typical notions but that they use a complex variety of co-existing and sometimes apparently contradictory norms to make sense of their situation. Basically similar is van Wijk's study of graphic designers which shows that identity construction is an ongoing dynamic process, which she traces from art college through employment in a design studio to debates about becoming a freelancer, in which the underlying value of creative freedom is framed in different contexts and identities.

These analyses do not support the thesis of identity in crisis, and at the same time amend the assumption of identification with 'the' organization by differentiating between various entities and goals a person can identify with and be committed to (cf. Arnold, Cooper and Robertson 1998, p. 211). Perhaps the idea of 'crisis' springs from a theoretical assumption about identity construction which does not recognize the ongoing dynamic and situational process of identity construction. Nevertheless, there seem to be situations characterized by ambiguity or discontinuity in which previous sense-making does not hold. Van Marrewijk discusses the situation of Dutch telecom employees who were recruited by an international alliance in which their own parent company participated together with three other national telecom operators. This international alliance had to operate in conditions which were vague and constantly changing because of shifting strategies of the four parent companies and in which experiences could not be shared by the employees because of differences in corporate and national cultures. As a consequence, employees shifted at times between a sense of belonging to their former colleagues in the Dutch parent company and distancing themselves from their former colleagues on the basis of seeing themselves as young, international, commercially orientated and dynamic, and likewise they felt attracted and repulsed by the international strategic alliance. In the process, organizational chaos and shifting identifications seem to have reinforced each other in a negative interaction, both at the level of the managers and employees of the international alliance and at the level of the top management of the parent companies. Arguably, employees felt frustrated by these shifting conditions which ended in the break-up of the international strategic alliance.

In the latter example, global competition in the telecom market and shifts in the power relations between different national operators determined the vicissitudes of organizational relationships in the international strategic alliance. Such macro conditions are even more prominent in the shifts in relationships between actors in the food production chain and have had a decisive influence on the position of farmers, about which Hennis and Ruivenkamp report. The techno-economic transformation of food production has affected the position of farmers and has made them increasingly dependent on suppliers of seeds and agrochemical pesticides on the one hand and on the food-processing industry on the other. However, the transformation has had a differentiated effect, thereby illustrating the room for farmers to impose their own definition of the situation, however small this room may be. Some farmers have had the means of growing into large-scale agro-businesses, have benefited from state price support and identify with entrepreneurship and with the food-processing multinationals. Other farmers have identified with a definition of farming which is related to preservation of the natural heritage and environment; they solicit financial support by the state and political coalitions with consumer and environmentalist organizations. In the process, traditional farmers' organizations have been unable to continue to claim that they are the general representative of farmers, as Hennis demonstrates, and new organizations have emerged mobilizing support for different identities of farmers. Ruivenkamp extends this analysis to include biotechnology researchers. Where traditional features of science, such as fundamental and applied research and publicly and privately funded research no longer divide because they have lost their meaning, researchers are invited to reflect on and redefine their position. The fact that some research institutes in India and Ghana have chosen to ally themselves with local sustainable development schemes marks the redefinition of their identity according to Ruivenkamp.

From the point of view of organizations and management it may be a cause of worry that employees in the case studies reported appear not to be committed to the organization as such, or to the top management. In fact, the lack of identification with the organization might raise the question what it is that prevents organizations from disintegrating. Several aspects of an answer to this question are provided by the contributors to this volume. Siebers and Vermeulen conclude that organizational bonding is created in the tension between management modes and the ways workers relate to them in specific practices. They use some strong wording to argue that the bonds that keep the organization together have a conditional and relational character, that bonding is about performance and strategy based on calculated and constructed interests: 'In social terms bonding is about power, in cultural terms it is about meanings and identifications being used to reach certain objectives. In our view an attempt to revive a romantic portrayal of bonding as a self-referential identification with the organization in trendy terms of "commitment", "loyalty" or "motivation" discloses a clear ideological character by ignoring these social and cultural dimensions.' Their contribution pays much attention to the various modes of management, which make particular demands on employees and managers and require certain levels of coordination. Through these coordination practices organizations are 'manufacturing consent' (Burawoy 1979) but such consent should not be understood as involving more than utilitarian acceptance on the side of employees. Another aspect of the bonds that sustain organizations comes from the chapter by Boessenkool, Leisink and Verweel, who regard organizational commitment as a double-edged sword. They argue that employees can be committed to their work or profession and identify with their colleagues and line manager without identifying with the organization as such, but managers, including top management, can relate to such commitments and identifications and make them instrumental for organizational purposes. This is not without risks, however, for organizational repertoires have an impact upon an individual's involvement in the organization, as Gastelaars shows in her contribution. Thus the culturing of organizational regimes, that is, the explicit catering to individual experiences, preferences and particularly to the symbolic values associated with the organization, can result in physical and/or psychological harassment, affecting one's attachment and loyalty towards an organization.

What do these findings say about the post-modernity cum organization theories from which several authors took their inspiration, and how do the contributors to this volume attempt to develop these theories? Several observations can be made, some of them reflecting on the presuppositions of the post-modernity theories themselves, and others elaborating the theoretical perspective on the dynamics of identity formation and bonding in the networking age. Grint explicitly addresses the post-modern supposition that the current situation is novel in the extent of its instability, and raises the question whether it matters for forms of organizational identity and binding whether the situation is unique or revolutionary or global. With respect to the former claim, he reflects that there is a tendency in each generation to see 'revolutions' that hindsight pronounces as merely marginal change, and argues that historically there have been many periods in which traditions were uprooted and whole societies either destabilized or even destroyed. It is the current perception of instability that has increased. With respect to the latter, Grint argues that 'there is no mechanical or deterministic link between "the situation" and the forms of identity because identity, as Anderson (1983) has insisted, lies in the imagination not in the so-called "real world"'. While Grint reflects critically on the historical and philosophical presuppositions of post-modern theories about detraditionalization and the crisis of identity, Boessenkool, Leisink and Verweel reflect on the situational fit between academic and management interest in organizational commitment and identification on the one hand and the (new) economy boom and the shortage in the labour market of the second half of the 1990s. Such management opportunism came out clearly when the new millennium brought economic recession and organizations started shedding labour and dropped the rhetoric of organizational identification and binding employees. To the opportunism of the 1990s' organizational discourse on social and cultural bonding they counterpose the conceptual differentiation between the mechanical rationality of productive organizations and the social rationality of living communities, and argue that the material wealth of the West makes it possible for organizations to transcend this dichotomy. However, Kamsteeg and Koot add some pertinent critical reflections. They observe that the options which management theory offers for redefining the situation by disidentification, paradoxical thinking, trust and associative logic are wasted on top managers, apart from a few exceptions. Most managers still cope by building their identity on performance and fall back on the exercise of power and control. More fundamentally it seems that Western management cannot relinquish its fixed belief in the possibilities of control, which is historically embedded in Enlightenment thinking.

Another important issue to reflect on concerns the question, what are adequate theoretical perspectives on the dynamics of identity formation and bonding in the networking age? This challenging question is taken up by Metz and Westenholz, whose purpose is to study the social construction of representations in terms of identities and situations that enable people to coordinate activities when they participate in so-called temporary and scattered work practices. Metz and Westenholz develop a relational perspective, arguing that identity is something that one develops in being recognized in relation to others, in relations in which recognition is negotiated. Using the empirical case of an IT programmer in a company, the theoretical perspective of communities of practice is shown to be useful for investigating identity in working life from a relational position. But since some negotiations of identities take place in loosely coupled situations rather than in communities of practice with established boundaries, the theoretical perspectives of knotworking and actor-network theory are introduced as supplementary perspectives which add insight into relations characterized as temporary and scattered, and into the role of technology in the process. The interpretive strength of the relational perspective which Metz and Westenholz illustrate in an empirical case analysis, shows the value of a theoretical perspective on identity construction which recognizes the ongoing dynamic and situational process of identity construction in which both actors – self and others – and organizational relationships and social structures are recursively involved.

DIRECTIONS FOR FUTURE RESEARCH

A common lesson from the empirical contributions to this volume is that employees in organizations are far more culturally adaptable than they are often assumed to be (see, for example, Beck 1992; Sennett 1998). Relevant examples in this book include Keenoy's academics, who have to deal with the erosion of their traditional academic freedom, as well as van Wijk's designers and van Marrewijk's telecom staff. In some cases, their 'solution' or coping strategy is not thoroughly thought-out in a logical sense, and there are – or so it seems – contradictions in their cultural repertoires. However, does this create a problem for anyone? Possibly it does for the (cognitive) researcher, who is trapped in his or her own pattern of standards and values of consistency and coherence, and consequently feels obliged to present a clear and internally non-contradictory picture of the strategies employed. But is it also a problem for the informant? It is likely that the informant will be satisfied simply if he or she succeeds in reducing his or her uncertainty and manages to devise a solution for his or her situation without creating any emotional problems, which is more probable in a cultural context where discontinuous behaviour is not taboo, such as in Western late modern society.

A question that follows naturally from this discussion is whether it would not be more sensible to devote more attention to culture and identity as power to survive and to reorganize, as urged by the (cognitive) anthropologists Turner (1993), Strauss and Quinn (1994) and Droogers (2001). The underlying idea is that people are able constantly to organize and reorganize their cultural knowledge because they use plans, which are minimal scripts for a given idea, emotion or action, often consisting of only a few characteristics. The plan is augmented and brought to life in each specific case (Droogers 2001, p.131). However, some plans are reasonably elaborate and have remained unchanged for a considerable time. Furthermore, some plans are linked to others to form a repertoire. Yet, people are capable of making repeated adjustments, even to such reasonably resilient plans. This is closely related to the fact that not all the available plans (or their properties) are used simultaneously, which provides for spare capacity. Furthermore, multiple plans can be followed at the same time, which is analogous to a large computer that derives its power from the fact that it actually consists of many small computers working in parallel. Finally, a repertoire can accommodate inconsistencies. All these considerations lead Turner (1993, p. 426) to ascribe a 'virtually infinite plasticity' to culture. This is not to say that culture implies nothing but dynamism. Continuity and discontinuity are both vital characteristics.

The same is true of identity, which can be described as an extension of culture as the human capacity to position oneself in a cultural field of power by making use of cultural plans and repertoires in a routine, yet often highly selective, way. We believe it is worth examining identity formation and binding in our late modern society from this cognitivist standpoint.

Moreover, it would be worthwhile to carry out a more comparative investigation into binding and identity construction in non-Western countries. The debate on organizational relations in the networking age focuses mainly on individuals in Western society, who in this late modern time are said to be trying to free themselves from the tentacles of the Enlightenment. In our part of the world, we are supposed to be experiencing a collision of modern and post-modern concepts of organization, identity formation and binding. However, similar processes are taking place in non-Western countries, where one can also observe that organizations and their employees (have to) survive and (learn to) cope with the tensions between old and new ideals (Clegg, Ibarra-Colado and Bueno-Rodriguez 1998). New Western (read, American) management models and organization concepts are currently also being introduced in Asian and African countries, where they have to be adapted to local cultures, which often have a completely different epistemological foundation (Coetzee and Roux 1998). For example, consider South Africa, where in common with Western countries we see (Knights and Willmott 1999; Koot and Sabelis 2002; Watson 1994) multiple, sometimes contradictory, simultaneous discourses on management and organization, sometimes in the same, but often in different contexts. The one (global) discourse is about new concepts such as self-management, chaos management, networking, the 'lean organization' and, at the same time, about traditional issues such as optimizing planning and efficiency (which produces a curious contradiction on its own), while the other (local) discourse is about 'African management' with such values as 'ubuntu' (collectiveness and concerted action), the here-and-now perception of time and spirituality. The first discourse is promoted mainly by South African management gurus trained in Europe and consultants from the large bureaus, while the second is promoted by politicians making a case for an 'African Renaissance', and organization experts urging localization (for example, Mbigi 2000). Some of the new management concepts to have come from the gurus logically fit in better with the basic assumptions of the black African cultures (such as 'the art of letting go' and the African perception of time), whereas others seem to be in conflict (such as self-management versus ubuntu). Cultural differentiation and fragmentation is increasing rapidly in South Africa too, partly as a consequence of processes such as individualization and migration. What is more, the discussions on the issue of cohesion and binding are just as lively there as here in the West. The question that still arises is whether black Africans cope with this cultural complexity in a different manner from Europeans and Americans, by which we mean coping in a cognitive, social, political and emotional sense.

The same applies to Asian countries, where Confucianism has produced a completely different view of the relationship with the world, namely one of adjustment and acceptance (cf. Landes 1998; Redding 1993; Weber 1951), which is about achieving as much self-control as possible. As it happens, this is exactly what the Western management gurus present as an ideal. Globalization is currently also under way in the Asian countries, and the new management concepts such as self-management are being introduced. How are they being adapted and hybridized into Confucianist thinking, and what does this mean for binding and identification of employees in Asian organizations?

Comparative investigation of this kind can lead not only to a further deepening, but also a validation or falsification of the insights into binding and identification that we have presented in this study.

Finally, we would like to remark that the empirical investigation and the theoretical insights presented in this book on organizational relationships in the networking age have led us to conclude that far more interdisciplinary or transdisciplinary research needs to be carried out. For example, the studies by van Marrewijk as well as those by Hennis and Ruivenkamp clearly show that it is necessary to view the issues of identity construction and bonding from a variety of disciplinary standpoints. There are psychological, anthropological, sociological, political, economic and technological aspects to consider and, furthermore, developments on macro, meso and micro levels are taking place alongside and in interaction with each other. In addition, Grint's contribution demonstrates the importance of a historical perspective.

The need for interdisciplinary research springs not only from the complex and dynamic object we are studying, but also from the related process of unbounding and pluralization of the human sciences themselves. In fact, the two processes of contemporary change are interrelated. On the one hand, there is a continuing process of unbounding and pluralization of the (macro and micro) socio-economic order in globalized societies, exemplified by the instabilities of transition in Eastern Europe and East Asia as well as in the everyday world of complex organizations and managerial work (for instance, Burton-Jones 1999; Castells 2000). Across both these macro and micro arenas, established structures have become increasingly compromised through processes of 'delayering', 'downsizing' and 'outsourcing', with knowledge work in organizational networks (multi-stakeholder, project-based and virtual) as the flourishing operational form (for example, Baumard 1999; Prichard et al. 2001). It is these changes which have prompted some theorists to hypothesize the 'crisis of identity'. On the other hand, there is the further unbounding and pluralization of the human sciences (see Geertz 1983; Gibbons et al. 1994), exemplified by paradigm shifts and emerging interdisciplinary arenas (for instance, the post-structural redefinition of 'subjects', territories and practices of knowledge), with intellectual networks (multi-stakeholder, project-based and virtual) as the flourishing operational form (see Caswill and Shove 2000; Nowotny, Scott and Gibbons 2001). Configured by competing understandings of the thresholds and boundaries of knowledge and of transition in organized life, the everyday world of organizational analysis has thus become increasingly compromised and hyperactive (see also Burrell 1997; Hassard and Parker 1993). Through these interrelated processes of unbounding and reshaping in both the theoretical and empirical foundations of the field, the mainstream of management and organizational knowledge has become both more pluralized and more performative.

These important and ongoing changes need to be understood in terms of broader changes that have been observed in the contemporary knowledge economy. An influential analysis of these matters has been provided by Gibbons et al. (1994), who argue that intensifying over the late 20th century has been a fundamental shift in knowledge relationships. Based on an investigation of the development of the humanities, natural and social sciences they argue that a new mode of knowledge production (Mode 2) has become overlaid upon a dominant mode of knowledge production (Mode 1) that was established earlier in modern society. They characterize Mode 1 as the conventional scientific model, which privileges the production of basic ('pure') knowledge within unified and concentrated mono-disciplinary subject areas. They characterize Mode 2 as more interactive, distributed and dynamic – arguing that in contemporary society the key repositories of knowledge and capability are distributed across conventional disciplinary and organizational boundaries, that is, those established under Mode 1. Furthermore, the effective networking of these resources would appear to enable more effective engagement with the complex problems of contemporary society, which have not tended to present themselves conveniently sectioned up in terms of the subject specialisms laid down in the modern university. Mode 2 is thus transdisciplinary: working beyond the boundaries of any one discipline; networked: socially distributed knowledge production sites that interconnect stakeholders who exchange perspectives and ideas; and interactive: knowledge is produced in the context of application through links between theory and practice.

Reviewing the ambitions of this book, it appears that we have navigated in the midst of this turbulence, both by the choice of our subject matter and by the attempt to combine several disciplinary perspectives and recognizing the constructed character of knowledge. In one sense we are pleased to present a critical demystifying analysis of the presumed turbulence of organizational relationships in the networking age and associated concepts of identity formation and bonding. In another sense we feel that we have only just begun to get a grip on the complexity of knowledge production in the everyday world of complex organizations and managerial work in late modern society.

REFERENCES

Arnold, J., C. L. Cooper and I. T. Robertson (1998), *Work Psychology: Understanding Human Behaviour in the Workplace*, third edition, Harlow: Pearson Education.

Ashforth, B. and S. Johnson (2001), 'Which hat to wear? The relative salience of multiple identities in organizational contexts', in A. Hogg and D. Terry (eds), *Social Identity Processes in Organizational Contexts*, Philadelphia: Psychology Press, pp. 31–49.

Baumard, P. (1999), *Tacit Knowledge in Organizations*, London: Sage.

Beck, U. (1992), *Risk Society: Towards a New Modernity*, London: Sage.

Burawoy, M. (1979), *Manufacturing Consent*, Chicago and London: University of Chicago Press.

Burrell, G. (1997), *Pandemonium: Towards a Retro-Organization Theory,* London: Sage.

Burton-Jones, A. (1999), *Knowledge Capitalism: Business, Work, and Learning in the New Economy*, Oxford: Oxford University Press.

Castells, M. (2000), *The Rise of the Network Society*, Oxford: Blackwell.

Clegg, S., E. Ibarra-Colado and L. Bueno-Rodriguez (1998), *Global Management, Universal Theories and Local Realities*, London: Sage.

Coetzee, P. and A. Roux (1998), *Philosophy from Africa*, Cape Town: Oxford University Press.

Droogers, A. (2001), 'Cultuur als repertoire', in E. Bartels et al. (eds), *Cultuur maken en cultuur breken*, Delft: Eburon, pp. 129–43.

Geertz, C. (1983), *Local Knowledge*, New York: Basic Books.
Gibbons, M., C. Limoges, H. Nowotny, S. Schwartzman, P. Scott and M. Trow (1994), *The New Production of Knowledge: The Dynamics of Science and Research in Contemporary Societies*, London: Sage.
Grint, K. (1997), *Fuzzy Management*, Oxford: Oxford University Press.
Hall, S. (1992), 'The question of cultural identity', in S. Hall, D. Held and T. McGrew (eds), *Modernity and its Futures*, Cambridge: Polity Press, pp. 273–316.
Hassard, J. and M. Parker (1993), *Postmodernism and Organisations*, London: Sage.
Heller, F., E. Pusic, G. Strauss and B. Wilpert (1998), *Organizational Participation: Myth and Reality*, Oxford: Oxford University Press.
Knights, D. and H. Willmott (1999), *Management Lives*, London: Sage.
Koot, W. and I. Sabelis (2002), *Beyond Complexity, Paradoxes and Coping Strategies in Managerial Life*, Amsterdam: Rozenberg.
Landes, D. (1998), *The Wealth and Poverty of Nations*, New York: Norton.
Lane, C. and R. Bachmann (eds) (2000), *Trust within and between Organizations*, Oxford: Oxford University Press.
Mbigi, L. (2000), *In Search of African Business Renaissance*, Randburg: Knowledge Resources.
Meyer, J. P. and N. J. Allen (1997), *Commitment in the Workplace*, Thousand Oaks, CA: Sage.
Nowotny, H., P. Scott and M. Gibbons (2001), *Rethinking Science, Knowledge and the Public in an Age of Uncertainty*, Cambridge: Polity Press.
Prichard, C., R. Hull, M. Chumer and H. Willmott (eds) (2000), *Managing Knowledge: Critical Investigations of Work and Learning*, London: Macmillan.
Redding, S. (1993), *The Spirit of Chinese Capitalism*, Berlin: De Gruyter.
Rousseau, D. (1995), *Psychological Contracts in Organizations: Understanding Written and Unwritten Agreements*, Thousand Oaks, CA: Sage.
Sennett, R. (1998), *The Corrosion of Character*, New York and London: Norton.
Strauss, C. and N. Quinn (1994), 'A cognitive anthropology', in R. Borofsky (ed.), *Assessing Cultural Anthropology*, New York: McGraw-Hill, pp. 284–97.
Thompson, K. (1992), 'Social pluralism and post-modernity', in S. Hall, D. Held and T. McGrew (eds), *Modernity and its Futures*, Cambridge: Polity Press, pp. 221–55.

Turner, T. (1993), 'Anthropology and multiculturalism: what is anthropology that multiculturalists should be mindful of it?', *Cultural Anthropology*, **8** (4), 411–29.
Watson, T. (1994), *In Search of Management*, London: Routledge.
Weber, M. (1951), *The Religion of China*, New York: Free Press.

Index